INTRODUCTION TO WinBUGS
FOR ECOLOGISTS

INTRODUCTION TO WinBUGS FOR ECOLOGISTS

A BAYESIAN APPROACH TO REGRESSION, ANOVA, MIXED MODELS, AND RELATED ANALYSES

MARC KÉRY
Swiss Ornithological Institute
6204 Sempach
Switzerland

AMSTERDAM • BOSTON • HEIDELBERG • LONDON
NEW YORK • OXFORD • PARIS • SAN DIEGO
SAN FRANCISCO • SINGAPORE • SYDNEY • TOKYO
Academic Press is an imprint of Elsevier

Academic Press is an imprint of Elsevier
30 Corporate Drive, Suite 400, Burlington, MA 01803, USA
525 B Street, Suite 1900, San Diego, CA 92101-4495, USA
84 Theobald's Road, London WC1X 8RR, UK
Radarweg 29, PO Box 211, 1000 AE Amsterdam, The Netherlands

First edition 2010

Notice
No responsibility is assumed by the publisher for any injury and/or damage to persons or property as a matter of products liability, negligence, or otherwise or from any use or operation of any methods, products, instructions, or ideas contained in the material herein. Because of rapid advances in the medical sciences, in particular, independent verification of diagnoses and drug dosages should be made.

Library of Congress Cataloging-in-Publication Data
Kéry, Marc.
  Introduction to WinBUGS for ecologists : A Bayesian approach to regression, ANOVA, mixed models and related analyses / Marc Kéry. – 1st ed.
    p. cm.
  ISBN: 978-0-12-378605-0
1. Biometry—Data processing. 2. WinBUGS. I. Title.
  QH323.5.K47 2010
  577.01'5118–dc22   2009048549

British Library Cataloguing in Publication Data
A catalogue record for this book is available from the British Library.
COVER PHOTOGRAPH: *Rosalia alpina* – T. Marent, Switzerland, 2008

For information on all Academic Press publications
visit our Web site at *www.books.elsevier.com*

Typeset by: diacriTech, Chennai, India

Printed and bound in China
11 12 13 14 15   10 9 8 7 6 5 4 3 2

Working together to grow
libraries in developing countries

www.elsevier.com | www.bookaid.org | www.sabre.org

ELSEVIER   **BOOK AID** International   Sabre Foundation

A Creed for Modeling

To make sense of an observation, everybody needs a model ...
whether he or she knows it or not.

It is difficult to imagine another method
that so effectively fosters clear thinking about a
system than the use of a model written in the language of algebra.

Contents

11. General Linear Model (ANCOVA)

12. Linear Mixed-Effects Model

13. Introduction to the Generalized Linear Model: Poisson "t-test"

14. Overdispersion, Zero-Inflation, and Offsets in the GLM

15. Poisson ANCOVA

16. Poisson Mixed-Effects Model (Poisson GLMM)

17. Binomial "t-Test"

18. Binomial Analysis of Covariance

19. Binomial Mixed-Effects Model (Binomial GLMM)

20. Nonstandard GLMMs 1: Site-Occupancy Species Distribution Model

21. Nonstandard GLMMs 2: Binomial Mixture Model to Model Abundance

22. Conclusions

Foreword

The title of Marc Kéry's book, *Introduction to WinBUGS for Ecologists*, provides some good hints about its content. From this title, we might guess that the book focuses on a piece of software, WinBUGS, that the treatment will not presuppose extensive knowledge of this software, and that the focus will be on the kinds of questions and inference problems that are faced by scientists who do ecology. So why WinBUGS and why ecologists? Of course, the most basic answer to this question is that Marc Kéry is an ecologist who has found WinBUGS to be extremely useful in his own work. But the important question then becomes, "Is Marc correct that WinBUGS can become an important tool for other ecologists?" The ultimate utility of this book will depend on the answer to this question, so I will try to develop a response here.

WinBUGS is a flexible, user-friendly software package that permits Bayesian inference from data, based on user-defined statistical models. Because the models must be completely specified by the user, WinBUGS may not be viewed by some as being as user-friendly as older statistical software packages that provide classical inference via methods such as maximum likelihood. So why should an ecologist invest the extra time and effort to learn WinBUGS? I can think of at least two reasons. The first is that all inference is based on underlying models (a basic constraint of the human condition). In the case of ecological data, the models represent caricatures of the processes that underlie both the data collection methods and the dynamics of ecological interest. I confess to knowing from personal experience that it is possible to obtain and "interpret" results of analyses from a standard statistical software package, without properly understanding the underlying model(s) on which inference was based. In contrast, having to specify a model in WinBUGS insures a basic understanding that need not accompany use of many common statistical software packages. So the necessity of specifying models, and thus of thinking clearly about underlying sampling and ecological processes, provide a good reason for ecologists to learn and use WinBUGS.

A second reason is that ecological data are typically generated by multiple processes, each of which induces variation. Frequently, such multiple sources of variation do not correspond closely to models available in more classical statistical software packages. Through my career as a quantitative ecologist, hundreds of field ecologists have brought me data sets asking for "standard" analyses, suggesting that I must have seen many similar data sets and that analysis of their data should thus be relatively quick and easy. However, despite these claims, I can't recall ever having seen a data set for which a standard, off-the-shelf analysis was strictly appropriate. There are always aspects of either the studied system or, more typically, the data collection process that requires nonstandard models. It was noted above that most ecological data sets are generated by at least two classes of process: ecological and sampling. The ecological process generates the true patterns that our studies are designed to investigate, and conditional on this truth, the sampling process generates the data that we actually obtain. The data are thus generated by multiple processes that are best viewed and modeled as hierarchical. Indeed, hierarchical models appropriate for such data are readily constructed in WinBUGS, and hierarchical Bayes provides a natural approach to inference for such data. Relative ease of implementation for complex hierarchical models is a compelling reason for ecologists to become proficient with WinBUGS.

For many problems, use of WinBUGS to implement a complex model can result in substantial savings of time and effort. I am always impressed by the small amount of WinBUGS code needed to provide inference for capture–recapture models that are represented by extremely complicated-looking likelihood functions. However, even more important than problems that can be solved more easily using WinBUGS than using traditional likelihood approaches are the problems that biologists would be unable to solve using these traditional approaches. For example, natural variation among individual organisms of a species will always exist for any characteristic under investigation. Such variation is pervasive and provides the raw material for Darwinian evolution by natural selection, the central guiding paradigm of all biological sciences. Even within groups of animals defined by age, sex, size, and other relevant covariates, ecologists still expect variation among individuals in virtually any attribute of interest. In capture–recapture modeling, for example, we would like to develop models capable of accounting for variation in capture probabilities and survival probabilities among individuals within any defined demographic group. However, in the absence of individual covariates, it is simply not possible to estimate a separate capture and survival probability for each individual animal. But we can consider a distribution of such probabilities across individuals and attempt to estimate characteristics of that distribution. In WinBUGS, we can develop hierarchical models in which among-individual variation is treated as a random effect, with this portion of the inference problem becoming one of estimating the parameters of the distributions that describe this individual variation. In contrast, I (and most ecologists) would not know how to begin to construct even moderately complex capture–recapture models with random individual effects using a likelihood framework. So WinBUGS provides access to models and inferences that would be otherwise unapproachable for most ecological scientists.

I conclude that WinBUGS, specifically, and hierarchical Bayesian analysis, generally, are probably very good things for ecologists to learn. However, we should still ask whether the book's contents and Marc's tutorial writing style are likely to provide readers with an adequate understanding of this material. My answer to this question is a resounding "Yes!" I especially like Marc's use of simulation to develop the data sets used in exercises and analyses throughout the book, as this approach effectively exploits the close connection between data analysis and generation. Statistical models are intended to be simplified representations of the processes that generate real data, and the repeated interplay between simulation and analysis provides an extremely effective means of teaching the ability to develop such models and understand the inferences that they produce.

Finally, I like the selection of models that are explored in this book. The bulk of the book focuses on general model classes that are used frequently by ecologists, as well as by scientists in other disciplines: linear models, generalized linear models, linear mixed models, and generalized linear mixed models. The learn-by-example approach of simulating data sets and analyzing them using both WinBUGS and classical approaches implemented in R provide an effective way not only to teach WinBUGS but also to provide a general understanding of these widely used classes of statistical model. The two chapters dealing with specific classes of ecological models, site occupancy models, and binomial mixture abundance models then provide the reader with an appreciation of the need to model both sampling and ecological processes in order to obtain reasonable inferences using data produced by actual ecological sampling. Indeed, it is in the development and application of models tailored to deal with specific ecological sampling methods that the power and utility of WinBUGS are most readily demonstrated.

However, I believe that the book, *Introduction to WinBUGS for Ecologists*, is far too modest and does not capture the central reasons why ecologists should read this book and work through the associated examples and exercises. Most important, I believe that the ecologist who gives this book a serious read will emerge with a good understanding of statistical models as abstract representations of the various processes that give rise to a data set. Such an understanding is basic to the development of inference models tailored to specific sampling and ecological scenarios. A benefit that will accompany this general understanding is specific insights into major classes of statistical models that are used in ecology and other areas of science. In addition, the tutorial development of models and analyses in WinBUGS and R should leave the reader with the ability to implement both standard and tailored models. I believe that it would be hard to over-state the value of adding to an ecologist's toolbox this ability to develop and then implement models tailored to specific studies.

Jim Nichols
Patuxent Wildlife Research Center, Laurel, MD

Preface

This book is a gentle introduction to applied Bayesian modeling for ecologists using the highly acclaimed, free WinBUGS software, as run from program R. The bulk of the book is formed by a very detailed yet, I hope, enjoyable tutorial consisting of commented example analyses. These form a progression from the trivially simple to the moderately complex and cover linear, generalized linear (GLM), mixed, and generalized linear mixed models (GLMMs). Along the way, a comprehensive and largely nonmathematical overview is given of these important model classes, which represent the core of modern applied statistics and are those which ecologists use most in their work. I provide complete R and WinBUGS code for all analyses; this allows you to follow them step-by-step and in the desired pace. Being an ecologist myself and having collaborated with many ecologist colleagues, I am convinced that the large majority of us best understands more complex statistical methods by first executing worked examples step-by-step and then by modifying these template analyses to fit their own data.

All analyses with WinBUGS are directly compared with analyses of the same data using standard R functions such as `lm()`, `glm()`, and `lmer()`. Hence, I would hope that this book will appeal to most ecologists regardless of whether they ultimately choose a Bayesian or a classical mode of inference for their analyses. In addition, the comparison of classical and Bayesian analyses should help demystify the Bayesian approach to statistical modeling. A key feature of this book is that all data sets are simulated (="assembled") before analysis (="disassembly") and that fully commented R code is provided for both. Data simulation, along with the powerful, yet intuitive model specification language in WinBUGS, represents a unique way to truly understand that core of applied statistics in much of ecology and other quantitative sciences, generalized linear models (GLMs) and mixed models.

This book traces my own journey as a quantitative ecologist toward an understanding of WinBUGS for Bayesian statistical modeling and of GLMs and mixed models. Both the simulation of data sets and model fitting in WinBUGS have been crucial for my own advancement in these respects. The book grew out of the documentation for a 1-week course that I teach at the graduate school for life sciences at the University of Zürich, Switzerland, and elsewhere to similar audiences. Therefore, the typical readership would be expected to be advanced undergraduate,

graduate students, and researchers in ecology and other quantitative sciences. To maximize your benefits, you should have some basic knowledge in R computing and statistics at the level of the linear model (LM) (i.e., analysis of variance and regression).

After three introductory chapters, normal LMs are dealt with in Chapters 4–11. In Chapter 9 and especially Chapter 12, they are generalized to contain more than a single stochastic process, i.e., to the (normal) linear mixed model (LMM). Chapter 13 introduces the GLM, i.e., the extension of the normal LM to allow error distributions other than the normal. Chapters 13–15 feature Poisson GLMs and Chapters 17–18 binomial GLMs. Finally, the GLM, too, is generalized to contain additional sources of random variation to become a GLMM in Chapter 16 for a Poisson example and in Chapter 19 for a binomial example. I strongly believe that this step-up approach, where the simplest of all LMs, that "of the mean" (Chapter 4), is made progressively more complex until we have a GLMM, helps you to get a synthetic understanding of these model classes, which have such a huge importance for applied statistics in ecology and elsewhere.

The final two main chapters go one step further and showcase two fairly novel and nonstandard versions of a GLMM. The first is the site-occupancy model for species distributions (Chapter 20; MacKenzie et al., 2002, 2003, 2006), and the second is the binomial (or N-) mixture model for estimation and modeling of abundance (Chapter 21; Royle, 2004). These models allow one to make inference about two pivotal quantities in ecology: distribution and abundance of a species (Krebs, 2001). Importantly, these models fully account for the imperfect detection of occupied sites and individuals, respectively. Arguably, imperfect detection is a hallmark of all ecological field studies. Hence, these models are extremely useful for ecologists but owing to their relative novelty are not yet widely known. Also, they are not usually described within the GLM framework, but I believe that recognizing how they fit into the larger picture of linear models is illuminating. The Bayesian analysis of these two models offers clear benefits over by that by maximum likelihood, for instance, in the ease with which finite-sample inference is obtained (Royle and Kéry, 2007), but also just heuristically, since these models are easier to understand when fit in WinBUGS.

Owing to its gentle tutorial style, this book should be excellent to teach yourself. I hope that you can learn much about Bayesian analysis using WinBUGS and about linear statistical models and their generalizations by simply reading it. However, the most effective way to do this obviously is by sitting at a computer and working through all examples, as well as by solving the exercises. Fairly often, I just give the code required to produce a certain output but do not show the actual result, so to fully grasp what is happening, it is best to execute all code.

If the book is used in a classroom setting and plenty of time is given to the solving of exercises, then up to two weeks might be required to cover all material. Alternatively, some chapters may be skipped or left for the students to go through for themselves. Chapters 1–5, inclusive, contain key material. If you already have experience with Bayesian inference, you may skip Chapters 1–2. If you understand well (generalized) linear models, you may also skip Chapter 6 and just skim Chapters 7–11 to see whether you can easily follow. Chapters 9 and 12 are the key chapters for your understanding of mixed models, whether LMM or GLMM, and should not be skipped. The same goes for Chapter 13, which introduces GLMs. The next Chapters (14–19) are examples of (mixed) GLMs and may be sampled selectively as desired. There is some redundancy in content, e.g., between the following pairs of chapters, which illustrate the same kind of model for a Poisson and a binomial response: 13/17, 15/18, and 16/19. Finally, Chapters 20 and 21 are somewhat more specialized and may not have the same importance for all readers (though I find them to be the most fascinating models in the whole book).

As much as I believe in the great benefits of data simulation for your understanding of a model, data assembly at the start of each chapter may be skipped. You can download all data sets from the book Web site or simply execute the R code to generate your own data sets and only go to the line-by-line mode of study where the analysis begins. Similarly, comparison of the Bayesian solutions with the maximum likelihood estimates can be dropped by simply fitting the models in WinBUGS and not in R also.

All R and WinBUGS code in this book can be downloaded from the book Web site at *http://www.mbr-pwrc.usgs.gov/software/kerybook/* maintained by Jim Hines at the Patuxent Wildlife Research Center. The Web site also contains some bonus material: a list of WinBUGS tricks, an Errata page, solutions to exercises, a text file containing all the code shown in the book, as well as the actual data sets that were used to produce the output shown in the book. It also contains a real data set (the Swiss hare data) we deal with extensively in the exercises. The Swiss hare data contain replicated counts of Brown hares (*Lepus europaeus*: see Chapter 13) conducted over 17 years (1992–2008) at 56 sites in eight regions of Switzerland. Replicated means that each year two counts were conducted during a 2-week period. Sites vary in area and elevation and belong to two types of habitat (arable and grassland): hence, there are both continuous and discrete explanatory variables. Unbounded counts may be modeled as Poisson random variables with *log(area)* as an offset, but we can also treat the observed density (i.e., the ratio of a count to area) as a normal or the incidence of a density exceeding some threshold as a binomial random variable. Hence, you can practice with all models shown in this book and meet many features of genuine data sets such as missing values and other nuisances of real life.

ACKNOWLEDGMENTS

My sincere thanks go to the following people: Andy Royle for introducing me to WinBUGS and for his friendship and plenty of help over the years; Jim Hines for creating and maintaining the book Web site, Michael (Miguel) Schaub, Benedikt Schmidt, Beth Gardner, and Bob Dorazio for help and discussions; Wesley Hochachka for reading and commenting on large parts of the book draft; the photographers for providing me with beautiful illustrations of the organisms behind our statistical examples, and Christophe Berney, Andrew Gelman, Jérôme Guélat, Fränzi Korner, Konstans Wells, and Ben Zuckerberg for various comments and help. Thanks also to the WinBUGS developers for making modern Bayesian statistical modeling accessible for nonstatisticians like myself and to the Swiss Ornithological Institute, especially Niklaus Zbinden and Lukas Jenni, for giving me some freedom to research and write. I am indebted to Jim Nichols for writing the Foreword to my book and for introducing me to hierarchical thinking in ecology during two extremely pleasant years spent at the Patuxent Wildlife Research Center. And finally, I express my deep gratitude to Susana and Gabriel for their love and patience.

Marc Kéry,
October 2009

1

Introduction

WinBUGS (Gilks et al., 1994; Spiegelhalter et al., 2003; Lunn et al., 2009) is a general-purpose software program to fit statistical models under the Bayesian approach to statistics. That is, statistical inference is based on the posterior distribution, which expresses all that is known about the parameters of a statistical model, given the data and existing knowledge. In recent years, the Bayesian paradigm has gained tremendous momentum

1

in statistics and its applications, including ecology, so it is natural to wonder about the reasons for this.

1.1 ADVANTAGES OF THE BAYESIAN APPROACH TO STATISTICS

Key assets of the Bayesian approach and of the associated computational methods include the following:

1.1.1 Numerical Tractability

Many statistical models are currently too complex to be fitted using classical statistical methods, but they can be fitted using Bayesian computational methods (Link et al., 2002). However, it may be reassuring that, in many cases, Bayesian inference gives answers that numerically closely match those obtained by classical methods.

1.1.2 Absence of Asymptotics

Asymptotically, that is, for a "large" sample, classical inference based on maximum likelihood (ML) is unbiased, i.e., in the long run right on target. However, for finite sample sizes, *i.e., for your data set*, ML may well be biased (Le Cam, 1990). Similarly, standard errors and confidence intervals are valid only for "large" samples. Statisticians never say what "large" exactly means, but you can be assured that typical ecological data sets aren't large. In contrast, Bayesian inference is *exact* for any sample size. This issue is not widely understood by ecological practitioners of statistics but may be particularly interesting for ecologists since our data sets are typically small to very small.

1.1.3 Ease of Error Propagation

In classical statistics, computing the uncertainty of functions of random variables such as parameters is not straightforward and involves approximations such as the delta method (Williams et al., 2002). For instance, consider obtaining an estimate for a population growth rate ($\hat{r}$) that is composed of two estimates of population size in subsequent years ($\hat{N}_1, \hat{N}_2$). We have $\hat{N}_1$ and $\hat{N}_2$ and we want $\hat{r}$: what should we do? Getting the point estimate of $\hat{r}$ is easy, but what about its standard error? In a Bayesian analysis with Markov chain Monte Carlo, estimating such, and much more complex, derived quantities including their uncertainty is trivial once we have a random sample from the posterior distribution of their constituent parts, such as $\hat{N}_1$ and $\hat{N}_2$ in our example.

1.1.4 Formal Framework for Combining Information

By basing inference on both what we knew before (the prior) and what we see now (the data at hand), and using solely the laws of probability for that combination, Bayesian statistics provides a formal mechanism for introducing external knowledge into an analysis. This may greatly increase the precision of the estimates (McCarthy and Masters, 2005); some parameters may only become estimable through precisely this combination of information.

Using existing information also appears a very sensible thing to do: after all, only rarely don't we know anything at all about the likely magnitude of an estimated parameter. For instance, when estimating the annual survival rate in a population of some large bird species such as a condor, we would be rather surprised to find it to be less than, say, 0.9. Values of less than, say, 0.5 would appear downright impossible. However, in classical statistics, by not using any existing information, we effectively say that the survival rate in that population could be just as well 0.1 as 0.9, or even 0 or 1. This is not really a sensible attitude since every population ecologists knows very well *a priori* that no condor population would ever survive for very long with a survival rate of 0.1. In classical statistics, we always feign total ignorance about the system under study when we analyze it.

However, within some limits, it is also possible to specify ignorance in a Bayesian analysis. That is, also under the Bayesian paradigm, we can base our inference on the observed data alone and thereby obtain inferences that are typically very similar numerically to those obtained in a classical analysis.

1.1.5 Intuitive Appeal

The *interpretation of probability* in the Bayesian paradigm is much more intuitive than in the classical statistical framework; in particular, we directly calculate the probability that a parameter has a certain value rather than the probability of obtaining a certain kind of data set, given some Null hypothesis. Hence, popular statements such as "I am 99% sure that ..." are only possible in a Bayesian mode of inference, but they are impossible in principle under the classical mode of inference. This is because, in the Bayesian approach, a probability statement is made about a parameter, whereas in the classical approach, it is about a data set.

Furthermore, by drawing conclusions based on a combination of what we knew before (the prior, or the "experience" part of learning) and what we see now (the likelihood, or the "current observation" part of learning), Bayesian statistics represent a *mathematical formalization of the learning process*, i.e., of how we all deal with and process information in science as well as in our daily life.

1.1.6 Coherence and Intellectual Beauty

The entire Bayesian theory of statistics is based on just three axioms of probability (Lindley, 1983, 2006). This contrasts with classical statistics that Bayesians are so fond to criticize for being a patchwork of theory and *ad hoc* amendments containing plenty of internal contradictions.

1.2 SO WHY THEN ISN'T EVERYONE A BAYESIAN?

Given all the advantages of the Bayesian approach to statistics just mentioned, it may come as a surprise that currently almost all ecologists still use classical statistics. Why is this?

Of course, there is some resistance to the Bayesian philosophy with its perceived subjectivity of prior choice and the challenge of avoiding to, unknowingly, inject information into an analysis via the priors, see Chapter 2. However, arguably, the lack of a much more widespread adoption of Bayesian methods in ecology has mostly practical reasons.

First, a Bayesian treatment shines most for complex models, which may not even be fit in a frequentist mode of inference (Link et al., 2002). Hence, until very recently, most applications of Bayesian statistics featured rather complex statistical models. These are neither the easiest to understand in the first place, nor may they be relevant to the majority of ecologists. Second, typical introductory books on Bayesian statistics are written in what is fairly heavy mathematics to most ecologists. Hence, getting to the entry point of the Bayesian world of statistics has been very difficult for many ecologists. Third, Bayesian philosophy and computational methods are not usually taught at universities. Finally, and perhaps most importantly, the practical implementation of a Bayesian analysis has typically involved custom-written code in general-purpose computer languages such as Fortran or C++. Therefore, for someone lacking a solid knowledge in statistics and computing, Bayesian analyses were essentially out of reach.

1.3 WinBUGS

This last point has radically changed with the advent of WinBUGS (Lunn et al., 2009). Arguably, WinBUGS is the only software that allows an average numerate ecologist to conduct his own Bayesian analyses of realistically complex, customized statistical models. By customized I mean that one is not constrained to run only those models that a program lets you select by clicking on a button. However, although WinBUGS has been and is increasingly being used in ecology, the paucity of really accessible and attractive introductions to WinBUGS for ecologists is a surprise (but see McCarthy, 2007). I believe that this is the main reason for why Win-BUGS isn't even more widely used in ecology.

1.4 WHY THIS BOOK?

This book aims at filling this gap by gently introducing ecologists to WinBUGS for exactly those methods they use most often, i.e., the linear, generalized linear, linear mixed, and generalized linear mixed model (GLMM). Table 1.1 shows how the three latter model classes are all generalizations of the simple Normal linear model (LM) in the top left cell of the body of the table. They extend the Normal model to contain either more than a single random process (represented by the residual in the Normal LM) and/or to exponential family distributions other than the Normal, e.g., Poisson and Binomial. Alternatively, starting from the GLMM in the bottom right cell, the other three model classes can be viewed as special cases obtained by imposing restrictions on a general GLMM.

These four model classes form the core of modern applied statistics. However, even though many ecologists will have applied them often using click-and-point programs or even statistics packages with a programming language such as GenStat, R, or SAS, I dare express doubts whether they all really always understand the models they have fitted. Having to specify a model in the elementary way that one has to in Win-BUGS will prove to greatly enhance your understanding of these models, whether you fit them by some sort of likelihood analysis (e.g., ML or restricted maximum likelihood [REML]) or in a Bayesian analysis.

Apart from the gentle and nonmathematical presentation by examples, the unique selling points of this book, which distinguish it from others, are

TABLE 1.1 Classification of Some Core Models Used for Applied Statistical Analysis

	Single Random Process	Two or More Random Processes
Normal response	Linear model (LM)	Linear mixed model (LMM)
Exponential family response	Generalized linear model (GLM)	Generalized linear mixed model (GLMM)

the full integration of all WinBUGS analyses into program R, the parallel presentation of classical and Bayesian analyses of all models and the use of simulated data sets. Next, I briefly expand on each of these points.

1.4.1 This Is Also an R Book

One key feature of this book as an introduction to WinBUGS is that we conduct all analyses in WinBUGS fully integrated within program R (R Development Core Team, 2007). R has become the *lingua franca* of modern statistical computing and conducting your Bayesian analysis in WinBUGS from within an R session has great practical benefits. Moreover, we also see how to conduct all analyses using common R functions such as `lm()`, `glm()`, and `glmer()`. This has the added bonus that this book will be useful to you even if you only want to learn to understand and fit the models in Table 1 in a classical statistical setting.

1.4.2 Juxtaposition of Classical and Bayesian Analyses

Another key feature is the juxtaposition of analyses using the classical methods provided for in program R (mostly ML) and the analyses of the same models in a Bayesian mode of inference using WinBUGS. Thus, with the exception of Chapters 20 and 21, we fit every model in both the classical and the Bayesian mode of inference. I have two reasons for creating parallel examples. First, this should increase your confidence into the "new" (Bayesian) solutions since with vague priors they give numerically very similar answers as the "old" solutions (e.g., ML). Second, the analysis of a single model by both classical and Bayesian methods should help to demystify Bayesian analysis. One sometimes reads statements like "we used a Bayesian model," or "perhaps a Bayesian model should be tried on this difficult problem." This is nonsense! Since any model exists independently of the method we choose to analyze it. For instance, the linear regression model is not Bayesian or non-Bayesian; rather, this model may be *analyzed* in a Bayesian or in a frequentist mode of inference. Even that class of models which has come to be seen as almost synonymous with Bayesian inference, hierarchical models which specify a hierarchy of stochastic processes, is not intrinsically Bayesian; rather, hierarchical models can be analyzed by frequentist (de Valpine and Hastings, 2002; Lee et al., 2006; de Valpine, 2009; Ponciano et al., 2009) or by Bayesian methods (Link and Sauer, 2002; Sauer and Link, 2002; Wikle, 2003; Clark et al., 2005). Indeed, many statisticians now use the two modes of inference quite opportunistically (Royle and Dorazio, 2006, 2008). Thus, the juxtaposition of classical and Bayesian analysis of the same models should make it very clear that a model is one thing and its analysis another and that there really is no such thing as a "Bayesian model."

1.4.3 The Power of Simulating Data

A third key feature of this book is the use of simulated data sets throughout (except for one data set used repeatedly in the exercises). At first, this may seem artificial, and I have no doubts that some readers may be disinterested in an analysis when a problem is perceived as "unreal." However, I would claim that several very important benefits accrue from the use of simulated data sets, especially in an introductory book:

1. For simulated data, truth is known. That is, estimates obtained in the analysis of a model can be compared with what we know they should be in the long-run average.
2. When coding an analysis in WinBUGS, especially in more complex cases but even for simpler ones, it is very easy to make mistakes. Ensuring that an analysis recovers estimates that resemble the known input values used to generate a data set can be an important check that it has been coded correctly.
3. It has been said that one of the most difficult, but absolutely necessary statistical concepts to grasp is that of the sampling variation of an estimator. For nonstatisticians, I don't see any other way to grasp the meaning of sampling variation other than literally experiencing it by repeatedly simulating data under the same model, analyzing them, and seeing how estimates differ randomly from one sample to the next: this variation is exactly what the standard error of an estimate quantifies. In real life, one typically only ever observes a single realization (i.e., data set) from the stochastic system about which one wants to make an inference in a statistical analysis. Hence, for ecologists it may be hard to make the connection with the concept of repeated samples from a system, when all we have is a single data set (and related to that, to understand the difference between a standard deviation and a standard error).
4. Simulating data can be used to study the long-run average characteristics of estimates, given a certain kind of data set, by repeating the same data generation-data analysis cycle many times. In this way, the (frequentist) operating characteristics of an estimator (bias, or "is it on target on average?"; efficiency, or "how far away from the target is the individual estimate on average?") can be studied by packaging both the simulation and the analysis into a loop and comparing the distribution of the resulting estimates to the known truth. Further, required sample sizes to obtain a desired level of precision can be investigated, as can issues of parameter estimability. All this can be done for exactly the specifications of one's data set, e.g., replicate data sets can be generated and analyzed with sample size and parameter values identical to those in one's real data set to get an impression, say, of the precision of the estimates that one is likely to

obtain. This is also the idea behind posterior predictive checks of goodness-of-fit, where the "natural" lack of fit for a model is studied using ideal data sets and then compared with the lack of fit observed for the actual data set (see Section 8.4.2).

5. Simulated data sets can be used to study effects of assumption violations. All models embody a set of assumptions that will be violated to some degree. Whether this has serious consequences for those estimates one is particularly interested in, can be studied using simulation.

6. Finally, and perhaps most importantly, I would claim that the ultimate proof that one has really understood the analysis of a statistical model is when one is able to simulate a data set under that very model. Analyzing data is a little like fixing a motorbike but in reverse: it consists of breaking a data set into its parts (e.g., covariate effects and variances), whereas fixing a bike means putting all the parts of a bike into the right place. One way to convince yourself that you really understand how a bike works is to first dismantle and then reassemble it again to a functioning vehicle. Similarly, for data analysis, by first assembling a data set and then breaking it apart into recognizable parts by analyzing it, you can prove to yourself that you really understand the analysis.

In summary, I believe that the value of simulation for analysis and understanding of complex stochastic systems can hardly be overstated. On a personal note, what has helped me most to understand nonnormal GLMs or mixed models, apart from having to specify them in the intuitive BUGS language, was to simulate the associated data sets in program R, which is great for simulating data.

Finally, I hope that the slightly artificial flavor of my data sets is more than made up for by their nice ecological setting and the attractive organisms we pretend to be studying. I imagine that many ecologists will by far prefer learning about new statistical methods using artificial *ecological* data sets than using real, but "boring" data sets from the political, social, economical, or medical sciences, as one has to do in many excellent introductory books.

1.5 WHAT THIS BOOK IS NOT ABOUT: THEORY OF BAYESIAN STATISTICS AND COMPUTATION

The theory of Bayesian inference is treated only very cursorily in this book (see Chapter 2). Other authors have done this admirably, and I refer you to them. Texts that should be accessible to ecologists include

Ellison (1996), Wade (2000), Link et al. (2002), Bernardo (2003), Brooks (2003), Gelman et al. (2004), Woodworth (2004), McCarthy (2007), Royle and Dorazio (2008), King et al. (2009), and Link and Barker (2010).

Furthermore, I don't dwell on explaining Markov chain Monte Carlo (MCMC) or Gibbs sampling, the computational methods most frequently used to fit models in the Bayesian framework. Arguably, a deep understanding of the details of MCMC is not required for an ecologist to conduct an adequate Bayesian analysis using WinBUGS. After all, very few ecologists who nowadays fit a GLM or a mixed model understand the (possibly restricted) likelihood function or the algorithms used to find its maximum. (Or can you explain the Newton–Raphson algorithm? And how about iteratively reweighted least squares?) Rather, by using WinBUGS we are going to experience some of the key features of MCMC. This includes the chain's initial transient behavior, the resultant need for visual or numerical assessment of convergence that leads to discarding of initial ("burn-in") parts of a chain, and the fact that successive iterations are not independent. If you want to read more on Bayesian computation, most of the above references may serve as an entry point to a rich literature.

1.6 FURTHER READING

If you seriously consider going Bayesian for your statistical modeling, you will probably want to purchase more than a single book. McCarthy (2007) is an accessible introduction to WinBUGS for beginners, although it presents WinBUGS only as a standalone application (i.e., not run from R) and the coverage of model classes dealt with is somewhat more limited. Gelman and Hill (2007) is an excellent textbook on linear, generalized, and mixed (generalized) linear models fit in both the classical and the Bayesian mode of inference and using both R and WinBUGS. Thus, its concept is somewhat similar to that of this book, though it does not feature the rigorous juxtaposition of both kinds of analysis. All examples are from the social and political sciences, which will perhaps not particularly interest an ecologist. However, the book contains a wealth of information that should be digestible for the audience of this book, as does Gelman et al. (2004). Ntzoufras (2009) is a new and comprehensive introduction to WinBUGS focusing on GLMs. It is very useful, but has a higher mathematical level and uses WinBUGS as a standalone application only. Woodworth (2004) is an entry-level introduction to Bayesian inference and also has some WinBUGS code examples.

Link and Barker (2010) is an excellent textbook on Bayesian inference specifically for ecologists and featuring numerous WinBUGS examples.

As an introduction to Bayesianism written mostly in everyday language, Lindley, an influential Bayesian thinker, has written a delightful book, where he argues, among others, that *probability is the extension of logic to all events, both certain (like classical logic) and uncertain* (Lindley, 2006, p. 66). His book is not about practical aspects of Bayesian analysis, but very informative, quite amusing and above all, written in an accessible way.

In this book, we run WinBUGS from within program R; hence, some knowledge of R is required. Your level of knowledge of R only needs to be minimal and any simple introduction to R would probably suffice to enable you to use this book. I like Dalgaard (2001) as a very accessible introduction that focuses mostly on linear models, and at a slightly higher level, featuring mostly GLMs, Crawley (2005) and Aitkin et al. (2009). More comprehensive R books will also contain everything required, e.g., Venables and Ripley (2002), Clark (2007), and Bolker (2008).

This book barely touches some of the statistical models that one would perhaps particularly expect to see in a statistics book for ecologists, namely, Chapters 20 and 21. I say nothing on such core topics in ecological statistics such as the estimation of population density, survival and other vital rates, or community parameters (Buckland et al., 2001; Borchers et al., 2002; Williams et al., 2002). This is intentional. I hope that my book lays the groundwork for a much better understanding of statistical modeling using WinBUGS. This will allow you to better tackle more complex and specialized analyses, including those featured in books like Royle and Dorazio (2008), King et al. (2009), and Link and Barker (2010).

Free documentation for WinBUGS abounds, see *http://www.mrc-bsu .cam.ac.uk/bugs/winbugs/contents.shtml*. The manual comes along with the program; within WinBUGS go `Help > User Manual` or press F1 and then scroll down. Recently, an open-source version of BUGS has been developed under the name of OpenBugs, see *http://mathstat.helsinki .fi/openbugs/*, and the latest release contains a set of ecological example analyses including those featured in Chapters 20 and 21. WinBUGS can be run in combination with other programs such as R, GenStat, Matlab, SAS; see the main WinBUGS Web site. There is even an Excel front-end (see *http://www.axrf86.dsl.pipex.com/*) that allows you to fit a wide range of complex models without even knowing the BUGS language. However, most serious WinBUGS users I know run it from R (see Chapter 5). It turns out that one of the main challenges for the budding WinBUGS programmer is to really understand the linear model (see Chapter 6). One particularly good introduction to the linear model in the context of survival and population estimation is Chapter 6 in Evan Cooch's *Gentle introduction to MARK* (see *http://www.phidot.org/software/mark/docs/book/pdf/chap6.pdf*).

1.7 SUMMARY

This book attempts the following:

1. *demystify Bayesian analyses* by showing their application in the most widely used general-purpose Bayesian software WinBUGS, in a gentle tutorial-like style and in parallel with classical analyses using program R, for a large set of ecological problems that range from very simple to moderately complex;
2. enhance your understanding of the *core of modern applied statistics*: linear, generalized linear, linear mixed, and generalized linear mixed models and features common to all of them, such as statistical distributions and the design matrix;
3. demonstrate the *great value of simulation*; and
4. thereby building a solid grounding of the use of WinBUGS (and R) for relatively simple models, so you can tackle more complex ones, and to help *free the modeler in you*.

Introduction to the Bayesian Analysis of a Statistical Model

This is a practical book that does not cover the theory of Bayesian analysis or computational methods in any detail. Nevertheless, a very brief introduction is in order. For more detailed expositions of these topics see: Royle and Dorazio (2008), King et al. (2009), or Link and Barker (2010), which are all specifically aimed at ecologists. In this chapter, I will first motivate a statistical view of ecology and the world at large and define a statistical model, then, contrast classical and Bayesian statistics, briefly touch upon Bayesian computation, sketch the steps of a typical Bayesian analysis, and finally, end with a brief pointer to special topics illustrated in this book.

2.1 PROBABILITY THEORY AND STATISTICS

Both probability theory and statistics are sciences that deal with uncertainty. Their subject is the description of stochastic systems, i.e., systems that are not fully predictable but include random processes that add a degree of chance—and therefore, uncertainty—in their outcome. Stochastic systems are ubiquitous in nature; hence, probability and statistics are important not only in science but also to understand all facets of life.

Indeed, stochastic systems are everywhere! They may be the weather ("will it rain tomorrow?"), politics ("will my party win?"), life ("will she marry me?"), sports ("will my football team win?"), an exam ("will I pass?"), the sex of an offspring ("will I have a daughter?"), body size of an organism, and *many, many more*. Indeed, it is hard to imagine anything observable in the world that is not at least in part affected by chance, i.e., at least partly unpredictable. For such observables or *data*, probability and statistics offer the only adequate framework for rigorous description, analysis, and prediction (Lindley, 2006).

To formally interpret any observation, we *always* need a model, i.e., an abstract description of how we believe our observations are a result of observable and unobservable quantities. The latter are called parameters, and one main aim of analyzing the model is to obtain numerical estimates for them. A model is always an abstraction and thus strictly always wrong. However, according to one of the most famous sayings in statistics, some models are useful and our goal must be to search for them. Useful models provide greater insights into a stochastic system that may otherwise be too complex to understand or to predict.

Both probability theory and statistics deal with the *characteristics of a stochastic system* (described by the parameters of a model) and its outcomes (the observed data), but these two fields represent different perspectives on stochastic systems. Probability theory specifies parameters and a model and then examines a variable outcome, whereas statistics takes the data, assumes a model, and then tries to infer the system properties, given the model. Parameters are key descriptors of the stochastic system about which one wants to learn something. Hence, statistics deals with making inferences (i.e., probabilistic conclusions about system components—parameters) based on a model and the observed outcome of a stochastic system.

2.2 TWO VIEWS OF STATISTICS: CLASSICAL AND BAYESIAN

In statistics, there are two main views about how one should learn about the parameter values in a stochastic system: classical (also called conventional or frequentist) and Bayesian statistics. Although practical applications of Bayesian statistics in ecology have greatly increased only in recent years, Bayesian statistics is, in fact, very old and was the dominating school of statistics for a long time. Indeed, the foundations of Bayesian statistics, the use of conditional probability for inference embodied in Bayes rule, were laid as early as 1763 by Thomas Bayes, an English minister and mathematician. In contrast, the foundations of classical statistics were not really laid until the first half of the twentieth century. So what are the differences?

Both classical and Bayesian statistics view data as the observed realizations of stochastic systems that contain one or several random processes. However, in classical statistics, the quantities used to describe these random processes (parameters) are fixed and unknown constants, whereas in Bayesian statistics, parameters are themselves viewed as unobserved realizations of random processes. In classical statistics, uncertainty is evaluated and described in terms of the frequency of hypothetical replicates, although these inferences are typically only described from knowledge of a single data set. Therefore, classical statistics is also called *frequentist statistics*. In Bayesian statistics, uncertainty is evaluated using the posterior distribution of a parameter, which is the conditional probability distribution of all unknown quantities (e.g., the parameters), given the data, the model, and what we knew about these quantities before conducting the analysis.

In other words, classical and Bayesian statistics differ in their definition of probability. In classical statistics, probability is the relative frequency of a feature of observed data. In contrast, in Bayesian statistics, probability is used to express one's uncertainty about the likely magnitude of a parameter; no hypothetical replication of the data set is required.

Under Bayesian inference, we fundamentally distinguish observable quantities x from unobservable quantities θ. Observables x are the data, whereas unobservables θ can be statistical parameters, missing data, mismeasured data, or future outcomes of the modeled system (predictions); they are all treated as random variables, i.e., quantities that can only be determined probabilistically. Because parameters θ are random variables under the Bayesian paradigm, we can make probabilistic statements about them, e.g., say things like "the probability that this population is in decline is 24%." In contrast, under the classical view of statistics, such statements are impossible in principle because parameters are fixed and only the data are random.

Central to both modes of inference is the sampling distribution $p(x | \theta)$ of the data x as a function of a model with its parameters θ. For instance, the sampling distribution for the total number of heads among 10 flips of a fair coin is the Binomial distribution with $p = 0.5$ and trial size $N = 10$. This is the distribution used to describe the effects of chance on the outcome of the random variable (here, sum of heads). In much of classical statistics, the likelihood function $p(x | \theta)$ is used as a basis for inference. The likelihood function is the same as the sampling distribution of the observed data x, but "read in the opposite direction": That value $\hat{\theta}$, which yields the maximum of the likelihood function for the observed data x is taken as the best estimate for θ and is called the *maximum likelihood estimate* (MLE) of the parameter θ. That is, much of classical inference is based on the estimation of a single point that corresponds to the maximum of a function. Note that θ can be a scalar or a vector.

The basis for Bayesian inference is Bayes rule, also called Bayes' theorem, which is a simple result of conditional probability. Bayes rule describes the relationship between the two conditional probabilities $p(A | B)$ and $p(B | A)$, where $|$ is read as "given":

$$p(A | B) = \frac{p(B | A)p(A)}{p(B)}.$$

This equation is an undisputed fact and can be proven from simple axioms of probability. However, what used to be more controversial, and partly still is (e.g., Dennis, 1996; de Valpine, 2009; Lele and Dennis, 2009; Ponciano et al., 2009), is how Bayes *used* his rule. He used it to derive the *probability of the parameters* θ, *given the data* x, that is, the *posterior distribution* $p(\theta | x)$:

$$p(\theta | x) = \frac{p(x | \theta)p(\theta)}{p(x)}.$$

We see that the posterior distribution $p(\theta | x)$ is proportional to the product of the likelihood function $p(x | \theta)$ and the prior distribution of the parameter $p(\theta)$. To make this product a genuine probability distribution function, with an integral equal to 1, a normalizing constant $p(x)$ is needed as a denominator; this is the probability of observing one's particular data set x. Ignoring the denominator (which is just a constant and does not involve the unknowns θ), Bayes, rule as applied in Bayesian statistics can be paraphrased as

Posterior distribution $\propto$ Likelihood $\times$ Prior distribution,

where $\propto$ reads as "is proportional to." Thus, Bayesian inference works by using the laws of probability to combine the information about parameter θ contained in the observed data x, as quantified in the likelihood function

$p(x|\theta)$, with what is known or assumed about the parameter before the data are collected or analyzed, i.e., the prior distribution $p(\theta)$. This results in a rigorous mathematical statement about the probability of parameter θ, given the data, the posterior *distribution* $p(\theta|x)$. Hence, while classical statistics works by estimating a single point for a parameter (which is an unknown constant), Bayesian statistics makes inference about an entire distribution instead, because parameters are random variables described by a statistical distribution.

A prior distribution does not necessarily imply a temporal priority, instead, it simply represents a specific assumption about a model parameter. Bayes rule tells us how to combine such an assumption about a parameter with our current observations into a logical, quantitative conclusion. The latter is represented by the posterior distribution of the parameter.

I find it hard not to be impressed by the application of Bayes rule to statistical inference, because it so *perfectly mimics the way in which we learn in everyday life!* In our guts, we always weigh any observation we make, or new information we get, with what we know to be the case or believe to know. For instance, if someone tells me that he went to the zoo and saw an elephant that stood 5 m tall, I believe this information and find the observation remarkable. However, if someone claimed that he just saw an elephant that stood 10 m tall, I don't believe him. This shows that human psychology works exactly as Bayes rule applied to statistical inference: we always weigh new information by its prior probability in drawing our conclusions (here, "Oh, that's amazing!" or "You must be mad!"). An elephant height of 10 m has a prior probability close to zero *to me*, hence, I am not all too impressed by this claim (except as to find it outrageous). Note, however, that I am not particularly knowledgeable about elephants. Perhaps someone with more specialist knowledge about pachyderms would already have serious doubts about the former claim (I haven't checked). This is the reason for why many Bayesians emphasize that all probability is subjective, or personal: it depends on what we knew before observing a datum (Lindley, 2006). It is easy to find plenty more examples of where we naturally *think* according to Bayes rule.

Inference in Bayesian statistics is a simple probability calculation, and one of the things Bayesians are most proud of is the parsimony and internal logic of their framework for inference. Thus, the entire Bayesian theory for inference can be derived using just three axioms of probability (Lindley, 1983, 2006). Bayes rule can be deduced from them, and the entire framework for Bayesian statistics, such as estimation, prediction, hypothesis testing, is based on just these three premises. In contrast, classical statistics lacks such an internally coherent body of theory.

However, the requirement to determine a prior probability $p(\theta)$ for model parameters ("prior belief") has caused fierce opposition to the

Bayesian paradigm because this was (and partly still is) seen to bring into science an unwanted subjective element. However, as we shall see, it is easy to exaggerate this issue, for several reasons. First, objective science or statistics is an illusion anyway: there are always decisions to be made, e.g., what questions to ask, what factor levels to study, whether to transform a response, and literally myriads more. Each one of these decisions may have an effect on the outcome of a study. Second, it is possible to use the Bayesian machinery for inference (Bayes rule and Markov chain Monte Carlo [MCMC] computing algorithms, see later) with so-called flat priors (also vague, diffuse, uninformative, minimally informative, or low-information priors). Such priors represent our ignorance about a parameter or our wish to let inference, i.e., the posterior distribution, be dominated by the observed data. Actually, this is exactly what we do throughout this book. Third, the prior is seen by some statisticians as a strength rather than a weakness of the Bayesian framework (Link and Barker, 2010): it lets one formally examine the effect on one's conclusions of different assumptions about the parameters. Also, anybody using informative priors must say so and justify this choice. When the choice of priors is suspected to have an undue influence on the posterior distribution, it is good practice to conduct a sensitivity analysis to see how much one's conclusions are changed when a different set of priors is used.

Nevertheless, it is fair to say that there can be challenges involving the priors. First, one possible problem is that priors are not invariant to transformation of parameters. A prior that is uninformative for θ may well be informative for a one-to-one transformation $g(\theta)$ of θ, such as $\log(\theta)$ or $1/\theta$. Hence, it is possible to introduce information into an analysis without intending to do so. Especially in complex models—and these are the ones where a Bayesian treatment and the Bayesian model fitting algorithms offer most rewards—it is quite possible that one unknowingly introduces unwanted information by the choice of ostensibly vague priors. Hence, for more complex models, a sensitivity analysis of priors is even more useful. Still, these challenges are not seen as insurmountable by many statisticians, and Bayesian statistics has now very much entered the mainstream of statistical science. This can be seen immediately when browsing journals such as *Biometrics*, *Biometrika*, or the *Journal of the American Statistical Association*, which contain both frequentist and Bayesian work. Many statisticians now use Bayesian and classical statistics alike, and some believe that in the future, we will see some kind of merging of the paradigms (e.g., Little, 2006).

Finally, in view of the profound philosophical difference between the two paradigms for statistical inference, it is remarkable how little parameter estimates actually differ numerically in practical applications when vague priors are used in a Bayesian analysis. We shall see this in almost every example in this book. Indeed, MLEs are an approximation to the mode of the posterior distribution of a parameter when vague priors are assumed. This is one of the reasons for the ironic claim made by I. J. Good "People who don't know they are Bayesians are called non-Bayesians."

2.3 THE IMPORTANCE OF MODERN ALGORITHMS AND COMPUTERS FOR BAYESIAN STATISTICS

For most modeling applications, the denominator in Bayes rule, $p(x)$, contains high-dimensional integrals which are analytically intractable. Historically, they had to be solved by more or less adequate numerical approximations. Often, they could not be solved at all. Ironically therefore, for a long time Bayesians thought that they had the better solutions in principle than classical statisticians but unfortunately could not practically apply them to any except very simple problems for want of a method to solve their equations.

A dramatic change of this situation came with the advent of simulation-based approaches like MCMC and related techniques that draw *samples from the posterior distribution* (see for instance the article entitled "Bayesian statistics without tears" by Smith and Gelfand, 1992). These techniques circumvent the need for actually computing the normalizing constant in Bayes rule. This, along with the ever-increasing computer power which is required for these highly iterative techniques, made the Bayesian revolution in statistics possible (Brooks, 2003).

It seems fair to say that the ease with which difficult computational problems are solved by MCMC algorithms is one of the main reasons for the recent upsurge of Bayesian statistics in ecology, rather than the ability to conduct an inference without pretending one is completely stupid (i.e., has no prior knowledge about the analyzed system). Indeed, so far there are only few articles in ecological journals that have actually used this asset of Bayesian statistics, i.e., have formally injected prior knowledge into their Bayesian analyses. They include Martin et al. (2005); McCarthy and Masters (2005); Mazzetta et al. (2007); and Swain et al. (2009). Nevertheless, it is likely that analyses with informative priors will become more common in the future.

2.4 MARKOV CHAIN MONTE CARLO (MCMC) AND GIBBS SAMPLING

MCMC is a set of techniques to simulate draws from the posterior distribution $p(\theta \,|\, x)$ given a model, a likelihood $p(x \,|\, \theta)$, and data x, using dependent sequences of random variables. That is, MCMC yields a sample from the posterior distribution of a parameter. MCMC was developed in 1953 by the physicists Metropolis et al., and later generalized by Hastings (1970), so one of the main MCMC algorithms is called the Metropolis–Hastings algorithm. Many different flavors of MCMC are available now. One of the most widely used MCMC techniques is Gibbs sampling (Geman and Geman, 1984). It is based on the idea that to solve a large problem, instead of trying to do all at once, it is more efficient to break the problem down into smaller subunits and solve each one in turn. Here is a sketch of how Gibbs sampling works taken from a course taught in 2005 by Nicky Best and Sylvia Richardson at Imperial College in London.

Let our data be x and our vector of unknowns θ consist of k subcomponents $\theta = (\theta_1, \theta_2, ..., \theta_k)$, hence we want to estimate k parameters.

1. Choose starting (initial) values $\quad \theta_1^{(0)}, \theta_2^{(0)}, ..., \theta_k^{(0)}$
2. Sample $\theta_1^{(1)}$ from $\qquad\qquad\qquad p(\theta_1 \mid \theta_2^{(0)}, \theta_3^{(0)}, ..., \theta_k^{(0)}, x)$

   Sample $\theta_2^{(1)}$ from $\qquad\qquad\qquad p(\theta_2 \mid \theta_1^{(1)}, \theta_3^{(0)}, ..., \theta_k^{(0)}, x)$

   ....

   Sample $\theta_k^{(1)}$ from $\qquad\qquad\qquad p(\theta_k \mid \theta_1^{(1)}, \theta_2^{(1)}, ..., \theta_{k-1}^{(1)}, x)$
3. Repeat step 2 many times (e.g. 100s, 1000s)
   —eventually obtain a sample from $p(\theta \mid x)$

Step 2 is called an update or iteration of the Gibbs sampler and after convergence is reached, it leads to one draw (=sample) consisting of k values from the joint posterior distribution $p(\theta \mid x)$. The conditional distributions in this step are called "full conditionals" as they condition on all other parameters. The sequence of random draws for each of k parameter resulting from step 3 forms a Markov chain.

So far, a very simplistic summary of a Bayesian statistical analysis as illustrated in this book would go as follows:

1. We use a degree-of-belief definition of probability rather than a definition of probability based on the frequency of events among hypothetical replicates.
2. We use probability distributions to summarize our beliefs or our knowledge (or lack thereof) about each model parameter and apply Bayes rule to update that knowledge with observed data to obtain the posterior distribution of every unknown in our model. The posterior distribution quantifies all our knowledge about these unknowns given the data, our model, and prior assumptions. All statistical inference is based on the posterior distribution.
3. However, posterior distributions are virtually impossible to compute analytically in all but the simplest cases; hence, we use simulation (MCMC) to draw series of dependent samples from the posterior distribution and base our inference on that sample.

How do we construct such a Gibbs sampler or other MCMC algorithm? I am told by statistician colleagues that this is surprisingly easy (but that the art consisted of constructing an *efficient* sampler). However, for most ecologists, this will arguably be prohibitively complicated. And this is where WinBUGS comes in: WinBUGS constructs an MCMC algorithm for us for the model specified and our data set and conducts the iterative simulations for as long as we desire and have time to wait. Essentially, WinBUGS is an MCMC blackbox (J. A. Royle, pers. comm.).

2.5 WHAT COMES AFTER MCMC?

Once we are done with MCMC, we have a series of random numbers from the joint posterior distribution $p(\theta | x)$ that may look like this for a two-parameter model such as the model of the mean in Chapters 4 and 5 (showing only the first six draws):

$$\mu: 4.28, 6.09, 7.37, 6.10, 4.72, 6.67, \ldots$$
$$\sigma^2: 10.98, 11.23, 15.26, 9.17, 14.82, 18.19, \ldots$$

Now what should we do with these numbers?

Essentially, we have to make sure that these numbers come from a stationary distribution, i.e., that the Markov chain that produced them was at an equilibrium. If that is the case, then this is our estimate of the posterior distribution. Also, these numbers should not be influenced by our choice of initial parameter values supplied to start the Markov chains (the initial values); remember that successive values are correlated. This is called convergence monitoring. Once we are satisfied, we can summarize our samples to estimate any desired feature of the posterior distribution we like, for instance, the mean, median, or mode as a measure of central tendency as a Bayesian point estimate or the standard deviation of the posterior distribution as a Bayesian measure of the uncertainty of a parameter estimate. Then, we can compute the posterior distribution for derived variables. For instance, if parameter γ is the ratio of α and β and we are interested in γ, we can simply divide α by β for each iteration in the Markov chain to obtain a sample of γ and then summarize that for inference about γ. We can also compute inferences for very complicated functions of parameters, such as the probability that γ exceeds some threshold value. Next, we briefly expand on each of these topics.

2.5.1 Convergence Monitoring

The first step in making an inference from an MCMC analysis is to ensure that an equilibrium distribution has indeed been reached by the

Markov chain, i.e., that the chain has *converged*. For each parameter, we started the chain at an arbitrary point (the initial value or *init* chosen for each parameter), and because successive draws are dependent on the previous values of each parameter, the actual values chosen for the inits will be noticeable for a while. Therefore, only after a while is the chain independent of the values with which it was started. These first draws ought to be discarded as a *burn-in* as they are unrepresentative of the equilibrium distribution of the Markov chain.

There are several ways to check for convergence. Most methods use at least two parallel chains, but another possibility is to compare successive sections of a single long chain. The simplest method is just to inspect plots of the chains visually: they should look like nice oscillograms around a horizontal line without any trend. Visual checks are routinely used to confirm convergence. For example, Fig. 2.1 shows the time-series plot for five parallel Markov chains for a parameter in a dynamic occupancy model (MacKenzie et al., 2003) fitted to 16 years worth of Swiss wallcreeper data (see Chapter 8). Convergence seems to be achieved after about 60 iterations.

Another, more formal check for convergence is based on the Gelman–Rubin (or Brooks–Gelman–Rubin) statistic (Gelman et al., 2004), called Rhat when using WinBUGS from R via R2WinBUGS (see Chapter 5). It compares between- and within-chain variance in an analysis of variance fashion. Values near 1 indicate likely convergence, and 1.1 is considered by some as an acceptable threshold (Gelman et al., 2004; Gelman and Hill, 2007). With this approach, it is important to start the parallel chains at different selected or at random places.

Convergence monitoring may be a thorny issue and there are horror stories about how difficult it can be to make sure that convergence has actually been achieved. I have repeatedly found cases where Rhat

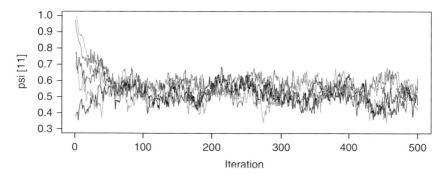

FIGURE 2.1   Time-series plot of five Markov chains for an occupancy parameter in a dynamic occupancy model fitted to Swiss wallcreeper data using WinBUGS (from Kéry et al., 2010a). Typically, the chains of all parameters do not converge equally rapidly.

erroneously indicated convergence; see Chapters 11 and 21 for examples. However, it is also easy to exaggerate this challenge, and with modern computing power, many models can be run for 100 000 iterations or more. Insuring convergence in MCMC analyses is in a sense akin to making sure that the global maximum in a likelihood function has been found in classical statistical analyses. In both, it can be difficult to determine that the desired goals have been achieved.

2.5.2 Summarizing the Posterior for Inference

Again, the aim of a Bayesian analysis is *not the estimate of a single point*, as the maximum of the likelihood function in classical statistics, but the estimate *of an entire distribution*. That means that every unknown (e.g., parameter, function of parameters, prediction, residual) has an entire distribution. This usually appears a bit odd at first. The posterior can be summarized graphically, e.g., using a histogram or a kernel-smoother. Alternatively, we can use mean, median, or mode as a measure of central tendency of a parameter (i.e., as a point estimate) and the standard deviation of the posterior as a measure of the uncertainty in the estimate, i.e., as the standard error of a parameter estimate. (Beware of challenging cases such as estimating a parameter that represents a standard deviation, e.g., the square root of a variance component. We will obtain the posterior distribution of that standard deviation, which will itself have a standard deviation that is used as the standard error of the estimate of the standard deviation ... simple, eh?). Finally, the Bayesian analog to a 95% confidence interval is called a (Bayesian) credible interval (CRI) and is any region of the posterior containing 95% of the area under the curve. There is more than one such region, and one particular CRI is the highest-posterior density interval (HPDI). However, in this book, we will only consider 95% CRIs bounded by the 2.5 and the 97.5 percentile points of the posterior sample of a parameter.

2.5.3 Computing Functions of Parameters

As mentioned earlier, one of the greatest features of the Bayesian mode of inference using MCMC is the ease with which *any function* of model parameters can be computed along with their standard errors exactly, while fully *accounting for all the uncertainty involved in computing this function* and without the need for any approximations such as the delta method (Powell, 2007). To get a posterior sample for a population growth rate r from two estimates of population size, N_1 and N_2, we simply compute the ratio of the two at every iteration of the Markov chain and summarize the resulting posterior sample for inference about r.

2.5.4 Forming Predictions

Predictions are expected values of the response for future samples or of hypothetical values of the explanatory variables of a model, or more generally, of any unobserved quantity. Predictions are very important for (a) presentation of results from an analysis and (b) to even understand what a model tells us. For example, the biological meaning of an interaction or a polynomial term can be difficult to determine from a set of parameter estimates. Because predictions are functions of parameters and of data (values of covariate), their posterior distributions can again be used for inference with the mean and the 95% CRIs often used as the predicted values along with a 95% prediction interval.

2.6 SOME SHARED CHALLENGES IN THE BAYESIAN AND THE CLASSICAL ANALYSIS OF A STATISTICAL MODEL

Other, much more general and also more difficult topics in a Bayesian analysis include model criticism (checking whether the chosen model is adequate for a data set), hypothesis testing and model selection, and checking parameter identifiability (e.g., making sure that there are enough data to actually estimate a parameter). These topics are also challenging in classical statistics, although they are frequently neglected.

2.6.1 Checking Model Adequacy

In models with a single error term (e.g., linear model [LM] and generalized linear model [GLM]), the usual residual diagnostics can be applied, e.g., plots of residuals vs. fitted values, histograms of the residuals, and so on, can be produced. We will see some examples of this in this book. In hierarchical models (i.e., models that include random effects other than residuals), checking model adequacy is more difficult and may have to involve (internal) cross-validation, validation against external data, or posterior predictive checks, see Gelman et al. (1996, 2004). We will see several examples for posterior predictive checks, including computation of the Bayesian p-value, which is not about hypothesis testing as in a classical analysis, but a measure of how well a model fits a given data set, i.e., a measure for goodness-of-fit.

2.6.2 Hypothesis Tests and Model Selection

As for classical statistics with a confidence interval, in the Bayesian paradigm, hypothesis tests can be conducted based on a CRI: if it overlaps zero, then the evidence for an effect of that parameter is ambiguous. In addition, we can make direct probability statements about the magnitude of a parameter as mentioned above.

To measure the evidence for a single or a collection of effects in the model, we can use an idea described by Kuo and Mallick (1998) and also Dellaportas et al. (2002): premultiply each regression parameter with a binary indicator w and give w a Bernoulli($p = 0.5$) prior. The posterior of w will then measure the probability that the associated effect belongs in the model; see Royle (2008) for an example of this. Based on the MCMC output from such a model run, model-averaged parameter estimates can also be produced. It must be noted, though, that implementing this feature greatly slows down MCMC samplers.

For model selection in nonhierarchical models, that is, models with a single random component, e.g., LMs and GLMs, there is a Bayesian analog to the Akaike's information criterion (AIC) called the DIC (deviance information criterion; Spiegelhalter et al., 2002). Similar to the AIC, the DIC is computed as the sum of the deviance plus twice the effective number of parameters (called pD) and expresses the trade-off between the fit of a model and the variance of (i.e., uncertainty around) its estimates. All else being equal, a more complex model fits better than a simpler one but has less precise parameter estimates, so the best choice will be some intermediate degree of model complexity.

The DIC as computed by WinBUGS seems to work well for nonhierarchical models, but unfortunately, for models more complex than GLMs, especially hierarchical models (e.g., linear mixed models and generalized linear mixed models), the DIC needs to be computed in a different and more complicated manner; see Millar (2009). Thus, in this book, we are not going to pay special attention to the DIC, nor to the estimate of the effective number of parameters (pD). However, you are invited to observe, for any of the models analyzed, whether pD or the DIC score computed make sense or not, for instance, when you add or drop a covariate or change other parts of a model.

There are other ways to decide on how much complexity is warranted in a model, one of which goes under the name reversible-jump (or RJ-) MCMC (King et al., 2009; Ntzoufras, 2009). Simple versions of RJ-MCMC can be implemented in WinBUGS, see *http://www.winbugs-development. org.uk/*.

No doubt the lack of a semiautomated way of conducting model selection or model averaging (except for RJ-MCMC) will come as a disappointment to many readers. After all, many ecologists have come to think that the problem of model selection has been solved and that this solution has a name, AIC (see extensive review by Burnham and Anderson, 2002). However, this rosy impression is probably too optimistic as argued, for instance, by Link and Barker (2006). It is perhaps instructive for an

TABLE 2.1   Examples for the Illustration of Some Important Special Topics

Topic	Location (Chapters)
Importance of detection probability when analyzing counts	13, 16, 16E, 17, 18, 20, and 21
Random effects/hierarchical models/mixed models	9, 12, 16, 19, 20, and 21
Computing residuals	7, 8, 13, 18, and 20
Posterior predictive checks, including Bayesian p-value	8, 13, 18, 20, and 21
Forming predictions	8, 10, 13, 15, 20, and 21
Prior sensitivity	5E, 20, and 21
Nonconvergence or convergence wrongly indicated	11 and 21
Missing values	4E and 8E
Standardization of covariates	11
Use of simulation for assessment of bias (or estimability) in an estimator	10

E denotes the exercises at the end of each chapter.

ecologist to browse through Kadane and Lazar (2004); this article published in one of the premier statistical research journals reviews model selection and clearly shows that in the field of statistics (as opposed to parts of ecology), the challenge of model selection is not yet viewed as having been solved.

2.6.3 Parameter Identifiability

A parameter can be loosely said to be identifiable when there is enough information in the data to determine its value unambiguously. This has nothing to do with the precision of the estimate. For instance, in the equation $a + b = 7$, no parameter is estimable, and we would need an additional equation in a and/or b to be able to determine the values of the two parameters, i.e., to make them estimable.

Strictly speaking, parameter identifiability is not an issue in the Bayesian framework because, in principle, we can always compute a posterior distribution for a parameter. At worst, the posterior will be the same as the prior, but then we haven't learned anything about that parameter. A common check for identifiability is therefore to compare the prior and the posterior distribution of a parameter: if the two coincide approximately, there does not seem to be any information in the data about that parameter. Assessing parameter identifiability is another difficult topic in the Bayesian world, and one where more research is needed, but the same state of affairs exists also for any complex statistical model analyzed by classical methods (see for instance Dennis et al., 2006). In WinBUGS, nonconvergence may not only indicate lack of identifiability of one or more parameters but also other problems. Perhaps one of the simplest ways of finding out whether a parameter is indeed identifiable is by simulation: we simulate a data set and see whether we are able to recover parameter values that resemble those used to generate the data. To distinguish sampling and estimation error from lack of estimability, simulations will have to be repeated many, e.g., 10 or 100, times.

2.7 POINTER TO SPECIAL TOPICS IN THIS BOOK

In this book, we learn extensively by example. Hence, Table 2.1. lists some special topics and shows where examples for them may be found.

2.8 SUMMARY

I have given a very brief introduction to Bayesian statistics and how it is conducted in practice using simulation-based methods (e.g., MCMC, Gibbs sampling). This chapter—and indeed the whole book—is *not* meant to deal with the theory of Bayesian statistics and the associated computational methods in any exhaustive way. Rather, books like King et al. (2009) and Link and Barker (2010) should be consulted for that.

EXERCISE

1. *Bayes rule in classical statistics*: Not every application of Bayes rule makes a probability calculation Bayesian as shown next in a classical example from medical testing. An important issue in medical testing is the probability that one actually has a disease (denoted "D"), given that one gets a positive test result, denoted "+" (which, depending on the test, in common language may be very negative, just think about an AIDS test). This probability is $p(D \mid +)$. With only three pieces of information that are often known for diagnostic tests and a given population, we can use Bayes rule to compute $p(D \mid +)$. We simply need to know the sensitivity of the diagnostic test, denoted $p(+ \mid D)$, its specificity $p(- \mid \text{not } D)$, and the general prevalence, or incidence, of the disease in the study population, $p(D)$. Note that sensitivity and specificity are the two possible kinds of diagnostic error.

Compute the probability of having the disease, given that you got a positive test result. Assume the following values: sensitivity = 0.99, specificity = 0.95, and prevalence = 5%. Start with $p(D\,|+) = \dfrac{p(+|D)p(D)}{p(+)}$ and note that a positive test result, which has probability $p(+)$, can be obtained in two ways: either one has the disease (with probability $p(D)$) or one does not have it (with probability $p(\text{not } D)$). Does the result surprise you?

3

WinBUGS

3.1 WHAT IS WinBUGS?

The BUGS language and program was developed by epidemiologists in Cambridge, UK in the 1990s (Gilks et al., 1994; Lunn et al., 2009). The acronym stands for **B**ayesian analysis **U**sing **G**ibbs **S**ampling. In later years, a Windows version called WinBUGS was developed (Spiegelhalter et al., 2003). Despite imperfections, (Win)BUGS is a groundbreaking program; for the first time, it has made really flexible and powerful Bayesian statistical modeling available to a large range of users, especially for users who lack the experience in statistics and computing to fit such fully custom models by maximizing their likelihood in a frequentist mode of inference. (Although no doubt some statisticians may deplore this because it may also lead to misuse; Lunn et al., 2009.)

WinBUGS lets one specify almost arbitrarily complex statistical models using a fairly simple model definition language that describes the stochastic and deterministic "local" relationships among all observable and unobservable quantities in a fully specified statistical model. These statistical models contain prior distributions for all top-level quantities, i.e., quantities that do not depend on other quantities. From this, WinBUGS determines the so-called full conditional distributions and then constructs a Gibbs or other MCMC sampler and uses it to produce the desired number of random samples from the joint posterior distribution.

Introduction to WinBUGS for Ecologists
DOI: 10.1016/B978-0-12-378605-0.00003-X

To let an ecologist really grasp what WinBUGS has brought us, it is instructive to compare the code (say in program R) required to find the maximum likelihood estimates (MLEs) of a custom model using numerical optimization with the code to specify the same model in WinBUGS: the difference is dramatic! With the former, some ugly likelihood expression appears at some point. In contrast, in WinBUGS, the likelihood of a model is specified implicitly as a series of deterministic and stochastic relationships; the latter types of relationships are specified as the statistical distributions assumed for all random quantities in the model.

WinBUGS is a fairly slow program; for large problems, it may fail to provide posterior samples of reasonable size within practical time limits. Custom-written samplers in more general programming languages such as Fortran or C++, or even R, can easily beat WinBUGS in terms of speed (Brooks, 2003), and often do so by a large margin. However, for many ecologists, writing their own samplers is simply not an option, and this makes WinBUGS so unique.

3.2 RUNNING WinBUGS FROM R

In contrast to most other WinBUGS introductions (e.g., McCarthy, 2007; Ntzoufras, 2009), all examples in this book will be analyzed with WinBUGS run from within program R by use of the R2WinBUGS package (Sturtz et al., 2005). R has become the *lingua franca* of statistical computing and conducting a Bayesian analysis in WinBUGS directly from within one's normal computing environment is a great advantage. In addition, the steps before and after an analysis in WinBUGS are greatly facilitated in R, e.g., data preparation as well as computations on the Markov chain Monte Carlo (MCMC) output and the presentation of results in graphs and tables.

Importantly, after conducting an analysis in WinBUGS, R2WinBUGS will import back into R the results of the Bayesian analysis, which essentially consist of the Markov chains for each monitored parameter. Hence, these results must be saved before exiting the program, otherwise all results will be lost! (Although the coda files are saved into the working directory, see 5.3.) This would not be dramatic for most examples in this book, but can be very annoying if you have just run a model for 7 days.

3.3 WinBUGS FREES THE MODELER IN YOU

Fitting statistical models in WinBUGS opens up a new world of modeling freedom to many ecologists. In my experience, writing WinBUGS code, or understanding and tailoring to one's own needs WinBUGS code written by others, is much more within the reach of typical ecologists than writing or adapting a similar analysis in some software that explicitly maximizes a likelihood function for the same problem. The simple pseudo-code model specification in WinBUGS is just wonderful. Thus, in theory, and often also in practice, WinBUGS really frees your creativity as a modeler and allows you to fit realistically complex models to observations from your study system.

However, it has been said (by Box or Cox, I believe) that statistical modeling is as much an art as a science. This applies particularly to modeling in WinBUGS, where for more complex problems, a lot of art (and patience!) may often be required to get an analysis running. WinBUGS is great when it works, but on the other hand, there are many things that may go wrong (e.g., traps, nonconvergence) without any obvious error in one's programming. For instance, cryptic error messages such as "trap 66" can make one feel really miserable. Of course, as usual, experience helps a great deal, so in an appendix on the book Web site I provide a list of tips that hopefully allow you to love WinBUGS more unconditionally (see *Appendix—A list of WinBUGS tricks*). I would suggest you skim over them now and then refresh your memory from time to time later.

3.4 SOME TECHNICALITIES AND CONVENTIONS

As a typographical convention in this book, WinBUGS and R code is shown in Courier font, `like this`. When part of the output was removed for the sake of clarity, I denote this by `[ ... ]`. When starting a new R session, it is useful or even obligate to set a few options, e.g., choose a working directory (where WinBUGS will save the files she creates, such as those containing the Markov chain values), load the R2WinBUGS package and tell R where the WinBUGS program is located. Each step may have to be adapted to your particular case. You can set the R working directory by issuing a command such as `setwd("C:\")`. The R2WinBUGS function `bugs()`, which we use all the time to call WinBUGS from within R, has an argument called `bugs.directory` that defaults to `"C:/Program Files/WinBUGS14/"` (though this is not the case under Windows VISTA). This is fine for most Anglosaxon computers but will have to be adapted for other language locales. Most recent versions of the `bugs()` function have another option called `working.directory` that partly overrides the global R setting from `setwd()`. It is best set to set `working.directory = getwd()`.

To conduct the analyses in this book, you will need to load some R packages, most notably R2WinBUGS, but sometimes also lme4, MASS, or lattice. At the beginning of every session, execute `library ("R2WinBUGS")` and do the same for other required packages. One useful R feature is a simple text file called `Rsite.profile`, which sits in `C:\Program Files\R\R-2.8.1\etc.`, and contains global R settings. For instance, you could add the following lines:

```
library("R2WinBUGS")
library("lme4")
bd <- "C:/Program files/WinBUGS14/"
```

This causes R to load the packages R2WinBUGS and lme4 whenever it is started and to have an object called `bd` that contains the usual WinBUGS address. We could add the option `bugs.directory = bd` whenever calling the `bugs()` function, and when running an analysis on a German-language locale, redefine `bd` appropriately.

In all analyses, I have striven for a consistent presentation. First, we write into the R working directory a text file containing the WinBUGS model description. After that, the other components required for the analysis are written into R objects: data, initial values, a list of parameters to be estimated and MCMC settings. Finally, these objects are sent to Win-BUGS by the R2WinBUGS workhorse function `bugs()`. Within the Win-BUGS model description, I have also sought consistency of layout by grouping as far as possible statements defining priors, the likelihood and derived quantities. This is not required; indeed, within some limits (e.g., putting code inside or outside of loops or pairs of braces), WinBUGS code lines may be swapped freely. This is not commonly the case with other programming languages.

When conducting a WinBUGS analysis from R, the WinBUGS model description, i.e., the text file that contains the model definition in the BUGS language, could be written in any text editor and then saved into the R working directory. However, I find it useful to use for that the `sink()` function and keep both WinBUGS and R code in the same file. However, you must not get confused about which code is R and which is WinBUGS: put simply, everything inside of the pair of `sink()` calls and inside the paired curly braces after the key word `model` will be WinBUGS code. The book Web site contains a text file with all the code shown in this book, and copying and pasting the code into an open R window works well. However, when using the popular R editor Tinn-R, a correct WinBUGS model description file will not always be produced when using `sink()`, so the WinBUGS trick list on the book Web site shows an alternative that works when using Tinn-R.

All analyses have been tested out with R 2.8.1 and WinBUGS 1.4.3. I would hope that some backwards compatibility is present. Also, most or all of the WinBUGS code should work fine in OpenBugs, but I have not verified this.

CHAPTER

4

A First Session in WinBUGS: The "Model of the Mean"

4.1 INTRODUCTION

To familiarize ourselves with WinBUGS and key features of a Bayesian analysis in practice (such as prior, likelihood, Markov chain Monte Carlo (MCMC) settings, initial values, updates, convergence, etc.), we will first run an analysis directly within WinBUGS of what is perhaps the simplest of all models—the "model of the mean." That is, we estimate the mean of a Normal population from a sample of measurements taken in that population. Our first example will deal with body mass of male peregrines (Fig. 4.1).

As an aside, this example emphasizes that even something that most people would not think of as a model—taking a simple average—in fact already *implies* a model that may be more or less adequate in any given case and is thus not as innocuous as many would think. For instance, using a simple average to express the central tendency of a skewed population is not very useful, and the "model of the mean" is then not adequate. Taking an average over nonindependent measurements, such as the diameter of each of several flowers measured within each of a sample

Introduction to WinBUGS for Ecologists
DOI: 10.1016/B978-0-12-378605-0.00004-1

FIGURE 4.1 Male peregrine falcon (*Falco peregrinus*) wintering in the French Mediterranean, Sète, 2008. (*Photo: J.-M. Delaunay*)

of plants, is also not such a good idea since the "model of the mean" underlying that average does not account for dependencies among measurements (as would, for instance, a mixed model). Finally, the computation and usual interpretation of a standard error for the population mean (as SD(x)/sqrt(n(x))) is implicitly based on a "model of the mean" where each measurement is independent and the population distribution reasonably close to a Normal or at least symmetric.

4.2 SETTING UP THE ANALYSIS

To open WinBUGS, we click on its bug-like icon. WinBUGS uses files in its own ODC format. To create an ODC file, you can either modify an existing ODC file (e.g., one of the examples contained in the WinBUGS help: go to Help > Examples vol I and II) or write a program in a text editor such as Word, save it as a text file (with ending .txt) and read it into WinBUGS and save it as an ODC document. Within WinBUGS, these files can be edited in content and format.

Here, we open the file called "The model of the (normal) mean.odc" (from the book Web site) via File > Open. We see something like at the top of the next page.

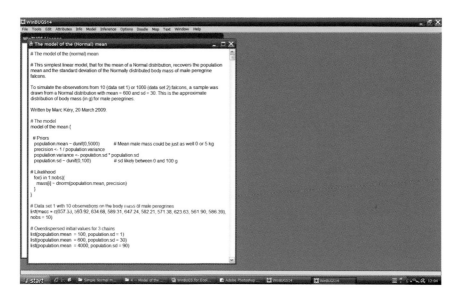

Note that we have two data sets available on the body mass of male peregrines. Both were created by using program R to produce random Normal numbers with specified values for the mean and the standard deviation. One data set contains 10 observations and the other contains 1000 (scroll down to see them). Male peregrines in Western Europe weigh on average about 600 g and Monneret (2006) gives a weight range of 500–680 g. Hence, the assumption of a Normal distribution of body mass implies a standard deviation of about 30 g.

To run a Bayesian analysis of the "model of the mean" for one of these data sets, we have to tell WinBUGS what the model is, what data to use, provide initial values from where to start the Markov chains and set some MCMC features, e.g., say how many Markov chains we want, how many draws from the posterior distribution we would like to have, how many draws WinBUGS should discard as a burn-in at the start and perhaps by

```
# Overdispersed initial values for 3 chains
list(population.mean  = 100, population.sd = 1)
list(population.mean  = 600, population.sd = 30)
list(population.mean  = 4000, population.sd = 90)
```

model is syntactically correct

how much we want to thin the resulting chain to save disk space and reduce autocorrelation among repeated draws.

First, we have to tell WinBUGS what the model is. We select Model > Specification, which causes the model Specification Tool window to pop up. Then, we put the cursor somewhere on the word "model" or indeed anywhere within the model definition in the ODC document and

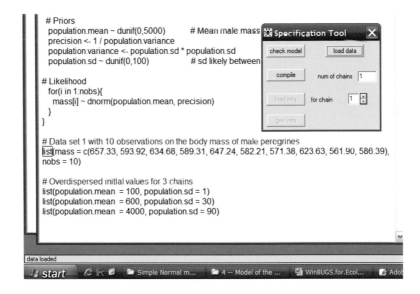

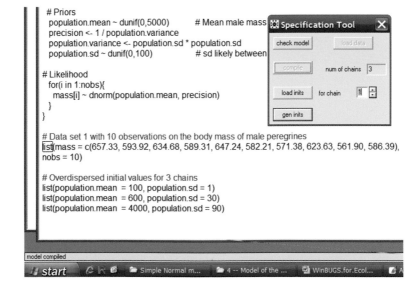

press "check model," whereupon, if the model is syntactically correct, WinBUGS responds in the bottom left corner:

Next, we load the data. We start with analysis of the data set number 1, which comprises 10 measurements, mark the entire word "list" in the data statement and press "load data", whereupon WinBUGS responds in the bottom left corner "data loaded."

Then, we select the number of parallel Markov chains that WinBUGS should "prepare" (i.e., compile), here, 3. This needs to be done *before* the "compile" button is pressed! Then press the "compile" button.

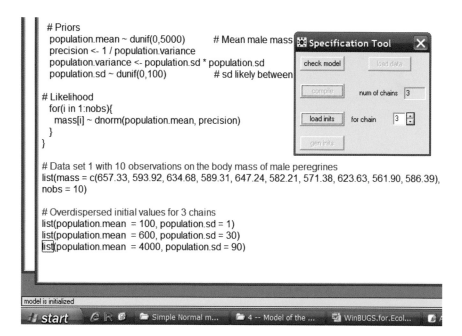

WinBUGS replies (in the bottom left corner) that the model is compiled. We still need to provide the starting values. One after another, mark the word "list" in that part of the program where the initial values are defined and then press "load inits." After you have done this once for each chain, the model is initialized and ready to run.

It is possible to let WinBUGS select the initial values by itself by instead pressing the "gen inits" button. This works often, and may be necessary for some parameters in more complex models, but in general, it is preferable to explicitly specify inits for as many parameters as possible.

Next, we choose Inference > Samples to tell WinBUGS what kind of inference we want for which parameter, which incidentally is called a *node* by WinBUGS. This makes the "Sample Monitor Tool" pop up (see top of p. 38).

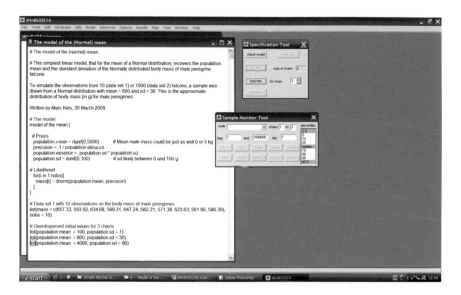

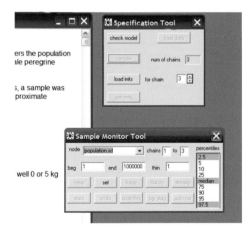

We write into the blank space next to the word "node" the names of those parameters whose posterior distributions we want to estimate by sampling them and press the "set" button. This button only becomes black, i.e., activated, once the correct name of a parameter from the previously defined model has been entered.

For models with many parameters, this can be tedious because we have to enter manually all of their names (correctly!). Moreover, every time we want to rerun the model, we have to repeat this whole procedure.

Once we have entered all parameters whose Markov chains we want to keep track of, i.e., for which we want to save the values and get estimates, we type an asterisk into the blank space to the right of the word "node."

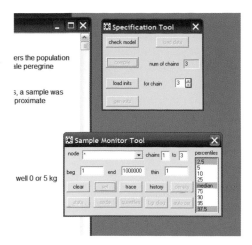

This tells WinBUGS that it should save the simulated draws from the posterior distributions of all the parameters whose names we have entered previously. On typing the asterisk, a few further buttons become activated, among them "trace" and "history" (see top of this page).

We select "trace," which will provide us with a moving time-series plot of the posterior draws for each selected parameter. (The button "history" produces a static graph of all draws up to the particular draw at which the history button is pushed.) Another window pops up that has an empty graph for every selected parameter. There are several other settings in this window that could be changed, perhaps only at later stages of the analysis. An important one is the "beg" and the "end" setting. Together, they define which draws should be used when summarizing the inference about each

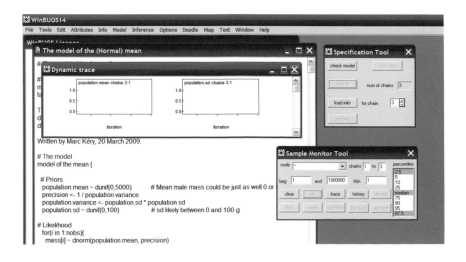

parameter, for instance when pressing the "density" or the "stats" buttons (see later). Changing the "beg" setting to, say, 100, specifies a burn-in of 100. This means that the first 100 draws from each Markov chain are discarded as not representative of the stationary distribution of the chain (i.e., the posterior distribution of the parameters in the model).

Now we are ready to start our simulation, i.e., start the Markov chains at their specified initial values and let the chains evolve for the requested number of times. To do this, we select Model > Update which makes a third window pop open, the "Update Tool."

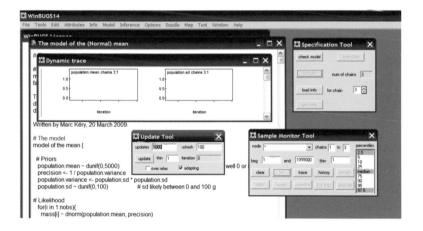

4.3 STARTING THE MCMC BLACKBOX

We choose the default 1000 updates, but change the "refresh" setting to 1 or 10, which gives a more smoothly moving dynamic trace of the Markov chains. Then, we press the "update" button and off we go. The

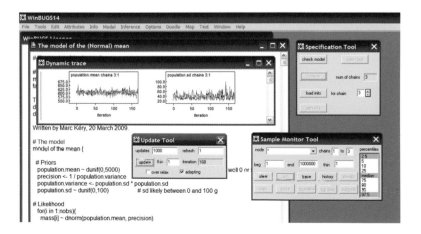

MCMC blackbox WinBUGS (Andy Royle, pers. comm.) draws samples from the posterior distributions of all monitored parameters. One continuous stream for each chain (here, 3) appears and represents the sampled values for each monitored parameter. Also, note that WinBUGS explains what it is doing at the bottom left corner, i.e., WinBUGS says that it is updating (=drawing samples from the posterior distribution).

Once the specified number of iterations (=updates, =draws) have been completed, WinBUGS stops and says that (on my laptop) the "updates took 62 s."

4.4 SUMMARIZING THE RESULTS

From each Markov chain, we now have a sample of 1000 random draws from the joint posterior distribution of the two parameters in the model. For inference about the mass of male peregrines, we can summarize these samples numerically or we can graph them, either in one dimension (for each parameter singly) or in two, to look at correlations among two parameters. If we wanted more samples, i.e., longer Markov chains, we could just press the "update" button again and would get another 1000 draws added. For now, let's be happy with a sample of 1000 (which is by far enough for this simple model and the small data set) and proceed with the inference about the parameters.

First we should check whether the Markov chains have indecd reached a stable equilibrium distribution, i.e., have *converged*. This can be done visually or by inspecting the Brooks–Gelman–Rubin (BGR) diagnostic statistic that WinBUGS displays on pressing the "bgr diag" button in the Sample Monitor Tool. The BGR statistic (graphed by the red line) is

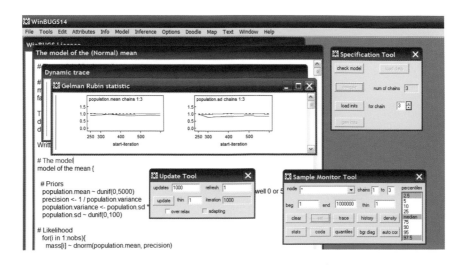

an analysis of variance (ANOVA)-type diagnostic that compares within- and among-chain variance. Values around 1 indicate convergence, with 1.1 considered as acceptable limit by Gelman and Hill (2007). Numerical summaries of the BGR statistic for successive sections of the chains can be obtained by first double-clicking on a BGR plot and then holding the CTRL key and clicking on the plot once again.

According to this diagnostic, the chains converge to a stationary distribution almost instantly; thus, we may well use all 3*1000 iterations for inference about the parameters. In contrast, for more complex models, nonzero burn-in lengths are always needed. For instance, for some complex models, some statisticians may use chains one million iterations long and routinely discard the first half of every chain as a burn-in.

Once we are satisfied that we have a valid sample from the posterior distribution of the parameters of the model of the mean and based on the observed 10 values of male peregrine body mass, we can use these draws to make an inference. Inference means to draw a probabilistic conclusion about the population from which these 10 peregrines came. Because we have created both this population and the samples from it, we know of course that the true population mean and standard deviations are 600 and 30 g, respectively.

We close some windows and press the "history," "density," "stats," and the "auto cor" buttons, which produces the following graphical display of information (after some rearranging of the pop-up windows):

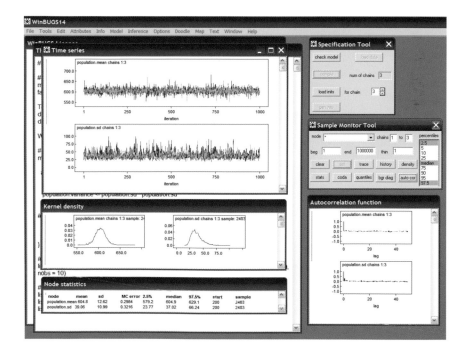

Visual inspection of the time series plot produced by "history" again suggests that the Markov chains have converged. Then, we obtain a kernel-smoothed histogram estimate of the posterior distributions of both parameters. The posterior of the mean looks symmetrical, while that for the standard deviation is fairly skewed. The node statistics give the formal parameter estimates. As a Bayesian point estimate, typically the posterior mean or the posterior median (or sometimes also the mode) is reported, while the posterior standard deviation is used as a standard error of the parameter estimate. The range between the 2.5th and 97.5th percentiles represents a 95% Bayesian confidence interval and is called a credible interval.

The autocorrelation function is depicted last. For this simple model, the chains are hardly autocorrelated at all. This is good as our posterior sample contains more information about the parameters than when successive draws are correlated. In more complex models, we will frequently find considerable autocorrelation between consecutive draws in a chain, and then it may be useful to thin the chains to get more approximately independent draws, i.e., more "concentrated" information. Note that a very high autocorrelation may also mean that a parameter is not identifiable in the model, i.e., not estimable, or that there are other structural problems with a model.

You should note two things about Bayesian credible intervals: First, though in practice, most nonstatisticians will not make a distinction between a Bayesian credible interval and a frequentist confidence interval, and indeed, for large samples sizes and vague priors, the two will typically be very similar numerically, there is in fact a major conceptual difference between them (see Chapter 2). Second, there are different ways in which the posterior distribution of a parameter may be summarized by a Bayesian credible interval. One is the highest posterior density interval (HPDI), which denotes the limits of the narrowest segment of the posterior distribution containing 95% of its total mass. HPDIs are not computed by WinBUGS (nor by R2WinBUGS, see later), but can be obtained using the function `HPDinterval()` in the R package coda.

One should also keep an eye on the Monte Carlo (MC) error under node statistics. The MC error quantifies the variability in the estimates that is due to Markov chain variability, which is the sampling error in the simulation-based solution for Bayes rule for our problem. MC error should be small. According to one rule of thumb, it should be <5% of the posterior standard deviation for a parameter.

One possible summary from our Bayesian analysis of the model of the mean adopted for the body mass of male peregrines would be that mean body mass is estimated at 604.8 g (SE: 12.62, 95%; CI: 579.2–629.1 g) and that the standard deviation of the body mass of male peregrines is 39.06 g (SE: 10.99, 95%; CI: 23.77–66.24 g). (Note how variance parameters are typically estimated much less precisely than are means.)

There is other summary information from WinBUGS that we can examine; for instance, Inference > Compare or Inference > Correlation allows us to plot parameter estimates or to see how they are related across two or more parameters. Let's have a look at whether the estimates of the mean body mass and its standard deviation are correlated: Choose Inference > Correlation, type the names of the two parameters (nodes) and we see that the estimates of the two parameters are not correlated at all.

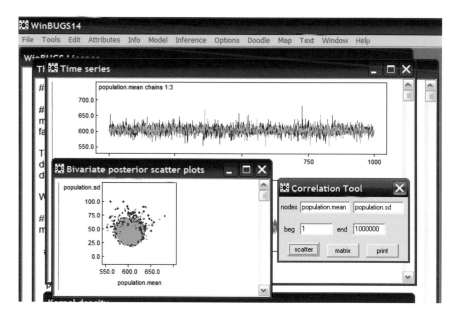

4.5 SUMMARY

We have conducted our first Bayesian analysis of the simplest statistical model, the "model of the mean." We see that WinBUGS is a very powerful software to fit models in a Bayesian mode of inference using MCMC and that a lot can be achieved using simple click and point techniques.

EXERCISES

1. Save your file under another name, e.g., test.odc. Then try out a few things that can go wrong.

   • create a typo in one of the node names of the model and WinBUGS complains "made use of undefined node."

- select extremely diffuse priors and see whether WinBUGS gets caught in a trap.
- select initial values outside of the range of a (uniform) prior and WinBUGS will crash.
- add data to the data set (e.g., `silly.dat = c(0,1,2)`) and WinBUGS will say "undefined variable."
- try out other typos in model, inits or data, e.g., an o ('oh') instead of a zero.
- See how WinBUGS deals with missing responses: turn the second element of the mass vector (in the smaller data set, but it doesn't really matter) into a missing value (NA). On loading the inits, you will now see that WinBUGS says (in its bottom left corner) that for every chain, there are still uninitialized variables. By this, it means the second element of mass, as we shall see. Continue (press the "gen inits" button) and then request a sample for both population.mean and mass in the Sample Monitor Tool. Request the Trace to be displayed and you will see that we get one for the node population.mean and another one for mass[2]. We see that this missing value is automatically imputed (estimated). The imputed value in this example is the same as the population mean, except that its uncertainty is much larger. Adding missing response values to the data set is one of the easiest ways in which predictions can be obtained in WinBUGS.

Running WinBUGS from R via R2WinBUGS

5.1 INTRODUCTION

In Chapter 4, it has become clear that as soon as you want to use WinBUGS more frequently, pointing and clicking is a pain and it is much more convenient to run the analyses by using a programming language. In program R, we can use the add-on package R2WinBUGS (Sturtz et al., 2005) to communicate with WinBUGS, so WinBUGS analyses can be run directly from R (also see package BRugs, not featured here, in connection with OpenBugs). The ingredients of an entire analysis are specified in R and sent to WinBUGS, then WinBUGS runs the desired number of updates, and the results, essentially just a long stream of random draws from the posterior distribution for each monitored parameter, along with an index showing which value of that stream belongs to which parameter, are sent back to R for convenient summary and further analysis. This is how we will be using WinBUGS from now on (so I will assume that you have R installed on your computer).

Any WinBUGS run produces a text file called "codaX.txt" for each chain X requested along with another text file called "codaIndex.txt."

When requesting three chains, we get `coda1.txt`, `coda2.txt`, `coda3.txt`, and the index file. They contain all sampled values of the Markov chains (MCs) and can be read into R using facilities provided by the R packages coda or boa and then analyzed further. Indeed, sometimes, for big models, the R object produced by R2WinBUGS may be too big for one's computer to swallow, although it may still be possible to read in the coda files. Doing this and using coda or boa for analyzing, the Markov chain Monte Carlo (MCMC) output may then be one's only way of obtaining inferences.

As an example for how to use WinBUGS in combination with R through R2WinBUGS, we rerun the model of the mean, this time for a custom-drawn sample of male peregrines. This is the first time where we use R code to simulate our data set and then analyze it in WinBUGS as well as by using the conventional facilities in R.

First, we assemble our data set, i.e., simulate a sample of male peregrine body mass measurements. Second, we use R to analyze the model of the mean for this sample in the frequentist mode of inference. Third, we use WinBUGS called from R to conduct the same analysis in a Bayesian mode of inference. Remember, analyzing data is like repairing motorcycles. Thus, analyzing the data set means breaking it apart into those pieces that we had used earlier to build it up. In real life, of course, nature assembles the data sets for us and we have to infer truth (i.e., nature's assembly rules).

5.2 DATA GENERATION

First, we create our data:

```
# Generate two samples of body mass measurements of male peregrines
y10 <- rnorm(n = 10, mean = 600, sd = 30)        # Sample of 10 birds
y1000 <- rnorm(n = 1000, mean = 600, sd = 30)   # Sample of 1000 birds

# Plot data
xlim = c(450, 750)
par(mfrow = c(2,1))
hist(y10, col = 'grey ', xlim = xlim, main = 'Body mass (g) of 10 male peregrines')
hist(y1000, col = 'grey', xlim = xlim, main = ' Body mass (g) of
1000 male peregrines')
```

Here, and indeed in all further examples, it is extremely instructive to execute the previous set of statements repeatedly to experience sampling error—the variation in one's data that stems from the fact that only part but not the whole of a variable population has been measured. Sampling error, or sampling variance, is something absolutely central to statistics, yet it is among the most difficult concepts for ecologists to grasp, especially, since in practice we only ever observe a single sample from the distribution that characterizes that variation! It is astonishing to observe how different repeated realizations from the exact same random process can be, here, the sampling of 10 or 1000 male peregrines from an assumed infinite population of male peregrines. Also surprising is, how far from normal the distribution in the smaller sample may look.

After playing around a while, we keep one pair of samples and go on. Note that unless you set a so-called seed for the random-number generator in R (for details, type ?set.seed), you will have a different sample from me (i.e., this book) and therefore also get slightly different output in the ensuing analyses (although you can download the exact data sets analyzed for the book from the Web site).

5.3 ANALYSIS USING R

We can conduct a quick classic analysis of this model using the linear regression facilities in R on the larger sample.

```
summary(lm(y1000 ~ 1))
> summary(lm(y1000 ~ 1))

[ ... ]

Coefficients:
             Estimate Std. Error t value Pr(>|t|)
(Intercept)  599.664      1.008   594.6   <2e-16 ***
---
Signif. codes:  0 '***' 0.001 '**' 0.01 '*' 0.05 '.' 0.1 ' ' 1

Residual standard error: 31.89 on 999 degrees of freedom
```

We recognize well the estimates of the population mean (599.664) and population standard deviation (SD), which is called the residual standard error (31.89). Now let's do the same analysis in WinBUGS.

5.4 ANALYSIS USING WinBUGS

Remember that you always need to load the R2WinBUGS package first, although we won't show this again from now on. Also, we need to set the R working directory.

```
library(R2WinBUGS)        # Load the R2WinBUGS library
setwd("F:/_WinBUGS book/Simple Normal mean model in WinBUGS") # wd

# Save BUGS description of the model to working directory
sink("model.txt")
```

```
cat("
model {

# Priors
 population.mean ~ dunif(0,5000)              # Normal parameterized by precision
 precision <- 1 / population.variance      # Precision = 1/variance
 population.variance <- population.sd * population.sd
 population.sd ~ dunif(0,100)

# Likelihood
 for(i in 1:nobs){
    mass[i] ~ dnorm(population.mean, precision)
 }
}
",fill=TRUE)
sink()
```

This last bit of code writes into the R working directory a text file named "model.txt" containing the WinBUGS description of the model. We can see this when we look at the Windows Explorer after execution of this set of statements.

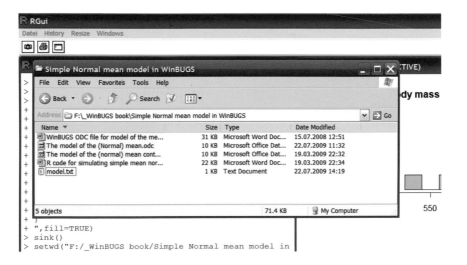

Next, we need to package the data that WinBUGS uses in the analysis. We do this by creating a bundle that contains both the data themselves and a count of the number of data points.

```
# Package all the stuff to be handed over to WinBUGS
# Bundle data
win.data <- list(mass = y1000, nobs = length(y1000))
```

Then, we define a function that creates random starting values, the inits function. We could also explicitly supply these initial values for each Markov chain requested, as we did in the previous chapter. However, this is less flexible, since we would need to explicitly specify one set of inits for each chain. If we wanted five chains instead of three, we would have to add two sets of inits. In contrast, a function will simply be executed two more times.

```
# Function to generate starting values
inits <- function()
  list(population.mean = rnorm(1,600), population.sd = runif(1,1,30))
```

As a reminder, if you are unsure what an R function such as `rnorm()` or `runif()` means, just type `?rnorm` or `?runif` in the R console. We also have to tell WinBUGS for which parameters it should save the posterior draws. Let's say we want to keep the variance also.

```
# Parameters to be monitored (= to estimate)
params <- c("population.mean", "population.sd", "population.variance")
```

Then, the MCMC settings need to be selected.

```
# MCMC settings
nc <- 3                 # Number of chains
ni <- 1000              # Number of draws from posterior (for each chain)
nb <- 1                 # Number of draws to discard as burn-in
nt <- 1                 # Thinning rate
```

Finally, the function `bugs()` is called to perform the analysis in WinBUGS and put its results into an R object called `out`:

```
# Start Gibbs sampler: Run model in WinBUGS and save results in object called out
out <- bugs(data = win.data, inits = inits, parameters.to.save = params, model.file
= "model.txt", n.thin = nt, n.chains = nc, n.burnin = nb, n.iter = ni, debug =
TRUE, DIC = TRUE, working.directory = getwd())
```

During execution of the program in WinBUGS, the R window is frozen. Once the requested number of draws from the posterior has been produced, Win-BUGS presents a graphical (the "history") and numerical summary of those parameters for which monitoring (i.e., estimation) was requested. On exiting WinBUGS, we have the results in various formats in our working directory (e.g., `coda1.txt`, `coda2.txt`, `coda3.txt`, `log.odc`, and `log.txt`) as well as in the R workspace a new object named `out`, a summary of which can be obtained by just typing its name (i.e., `out`). Note that setting debug = FALSE would cause WinBUGS to exit automatically after completion of the requested number of draws. This is important for instance when running repeated simulations. Otherwise, keeping WinBUGS open after the required number of draws have been taken from the posterior (i.e., setting debug = TRUE) allows you to use WinBUGS for additional analyses, for instance, to make plots.

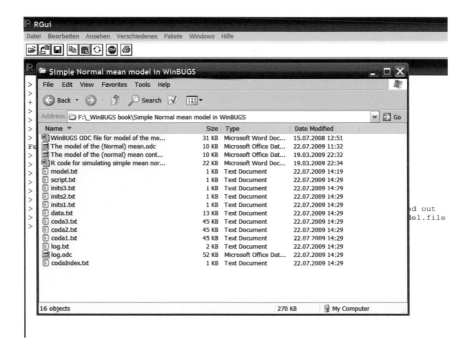

Now look at the R workspace and note the new object called *out*:

```
ls()
> ls()
 [1] "inits"     "nb"     "nc"     "ni"     "nt"    "out"    "params"
 [8] "win.data"  "xlim"   "y10"    "y1000"
```

We look at a summary of the Bayesian analysis:

```
out       # Produces a summary of the object
> out
Inference for Bugs model at "model.txt", fit using WinBUGS,
  3 chains, each with 1000 iterations (first 1 discarded)
  n.sims = 2997 iterations saved
```

	mean	sd	2.5%	25%	50%	75%	97.5%	Rhat	n.eff
population.mean	599.7	1.0	597.7	598.9	599.6	600.4	601.7	1.0	3000
population.sd	32.0	1.8	30.6	31.4	31.9	32.4	33.4	1.1	3000
population.variance	1027.0	173.7	934.6	988.7	1019.0	1050.0	1116.0	1.1	3000
deviance	9765.1	33.4	9762.0	9762.0	9763.0	9764.0	9769.0	1.1	1500

For each parameter, n.eff is a crude measure of effective sample size,
and Rhat is the potential scale reduction factor (at convergence, Rhat=1).

DIC info (using the rule, pD = Dbar-Dhat)
pD = 3.5 and DIC = 9768.6
DIC is an estimate of expected predictive error (lower deviance is better).

Actually, object *out* contains a lot more information, which we can see by
listing all the objects contained within *out* by typing `names(out)`.

```
> names(out)
 [1] "n.chains"       "n.iter"        "n.burnin"       "n.thin"
 [5] "n.keep"         "n.sims"        "sims.array"     "sims.list"
 [9] "sims.matrix"    "summary"       "mean"           "sd"
[13] "median"         "root.short"    "long.short"     "dimension.short"
[17] "indexes.short"  "last.values"   "isDIC"          "DICbyR"
[21] "pD"             "DIC"           "model.file"     "program"
```

You can look into any of these objects by typing their name, which will
usually print their full content, or by applying some summarizing function
such as `names()` or `str()` on them. (Again, when not sure what an R function does, just type `?names` or `?str` into your R console.) You can also look
deeper; note some degree of redundancy among `sims.array`, `sims.list`,
and `sims.matrix`:

```
str(out)
> str(out)
List of 24
 $ n.chains     : num 3
 $ n.iter       : num 1000
 $ n.burnin     : num 1
 $ n.thin       : num 1
 $ n.keep       : num 999
 $ n.sims       : num 2997
 $ sims.array   : num [1:999, 1:3, 1:4] 600 601 600 599 599 ...
  ..- attr(*, "dimnames")=List of 3
  .. ..$ : NULL
  .. ..$ : NULL
  .. ..$ : chr [1:4] "population.mean" "population.sd" "population.variance"
"deviance"
 $ sims.list    :List of 4
  ..$ population.mean    : num [1:2997] 600 600 601 601 600 ...
  ..$ population.sd      : num [1:2997] 31.8 32.8 31.6 32 31.8 ...
  ..$ population.variance: num [1:2997] 1013 1076 1000 1027 1011 ...
  ..$ deviance           : num [1:2997] 9762 9763 9764 9763 9762 ...
 $ sims.matrix  : num [1:2997, 1:4] 600 600 601 601 600 ...
```

```
    ..- attr(*, "dimnames")=List of 2
    .. ..$ : NULL
    .. ..$ : chr [1:4] "population.mean" "population.sd" "population.variance"
  "deviance"
   $ summary        : num [1:4, 1:9] 599.66 32 1027.02 9765.13 1.04 ...
    ..- attr(*, "dimnames")=List of 2
    .. ..$ : chr [1:4] "population.mean" "population.sd" "population.variance"
  "deviance"
    .. ..$ : chr [1:9] "mean" "sd" "2.5%" "25%" ...
   $ mean            :List of 4
   ..$ population.mean     : num 600
   ..$ population.sd       : num 32
   ..$ population.variance: num 1027
   ..$ deviance            : num 9765
   $ sd              :List of 4
   ..$ population.mean     : num 1.04
   ..$ population.sd       : num 1.78
   ..$ population.variance: num 174
   ..$ deviance            : num 33.4
   $ median          :List of 4
   ..$ population.mean     : num 600
   ..$ population.sd       : num 31.9
   ..$ population.variance: num 1019
   ..$ deviance            : num 9763
  [ ... ]
```

Object *out* contains all the information contained in the coda files that WinBUGS produced plus some processed items such as summaries like mean values of all monitored parameters, the Brooks–Gelman–Rubin (BGR) convergence diagnostic called Rhat, and an effective sample size that corrects for the degree of autocorrelation within the chains (more autocorrelation means smaller effective sample size). We can now apply standard R commands to get what we want from this raw output of our Bayesian analysis of the model of the mean.

For a quick check whether any of the parameters has a BGR diagnostic greater than 1.1 (i.e., has Markov chains that have not converged), you can type this:

```
hist(out$summary[,8])        # Rhat values in the eighth column of the summary
which(out$summary[,8] > 1.1) # None in this case
```

For trace plots for the entire chains, do

```
par(mfrow = c(3,1))
matplot(out$sims.array[1:999,1:3,1], type = "l")
matplot(out$sims.array[,,2] , type = "l")
matplot(out$sims.array[,,3] , type = "l")
```

... or just for the start of the chains to see how rapidly they converge ...

```
par(mfrow = c(3,1))
matplot(out$sims.array[1:20,1:3,1], type = "l")
matplot(out$sims.array[1:20,,2] , type = "l")
matplot(out$sims.array[1:20,,3] , type = "l")
```

We can also produce graphical summaries, e.g., histograms of the posterior distributions for each parameter:

```
par(mfrow = c(3,1))
hist(out$sims.list$population.mean, col = "grey")
hist(out$sims.list$population.sd, col = "blue")
hist(out$sims.list$population.variance, col = "green")
```

... or plot the (lack of) correlation between two parameters:

```
par(mfrow = c(1,1))
plot(out$sims.list$population.mean, out$sims.list$population.sd)
```

or

```
pairs(cbind(out$sims.list$population.mean, out$sims.list$population.sd,
out$sims.list$population.variance))
```

Numerical summaries of the posterior distribution can also be obtained, with the standard deviation requested separately:

```
summary(out$sims.list$population.mean)
summary(out$sims.list$population.sd)
sd(out$sims.list$population.mean)
sd(out$sims.list$population.sd)
```

Now compare this again with the classical analysis using maximum likelihood and the R function lm(). The results are almost indistinguishable.

```
summary(lm(y1000 ~ 1))
```

To summarize, after using R2WinBUGS, the R workspace contains all the results from your Bayesian analysis (the coda files contain them as well but in a less accessible format). If you want to keep these results, you must save the R workspace or save them in an external file.

5.5 SUMMARY

We have repeated the analysis of the model of the mean from the previous chapter by calling WinBUGS from within program R using the R2WinBUGS interface package. This is virtually always the most efficient mode of running analyses in WinBUGS and is the way in which we will use WinBUGS in the remainder of this book. We have also seen that a

Bayesian analysis with vague priors yields almost identical estimates as a classical analysis, something that we will see many more times.

EXERCISES

1. *Informative priors*: The analysis just presented uses vague priors, i.e., the effect of the prior on the posterior distribution is minimal. As we have just seen, Bayesian analyses with such priors typically yield estimates very similar numerically to those obtained using maximum likelihood.

   To see how the posterior distribution is influenced by both prior and likelihood, assume that we knew that the average body mass of male peregrines in populations like the one we sampled lay between 500 and 590 g. We can then formally incorporate this information into the analysis by means of the prior distribution. Hint: Change some code bits as follows, and let WinBUGS generate initial values automatically:

   ```
   # Priors
   population.mean ~ dunif(500, 590) # Mean mass must lie between 500 and 590
   ...
   ```

   Rerun the analysis (again, with the prior for the `population.sd` slightly changed) and see how the parameter estimates and their uncertainty change under this prior. Is this a bad thing? You may want to experiment with other limits for the uniform prior on the sd, for instance, set it at (0,10). Run the model and inspect the posterior distributions.
2. Run the model for the small and large data set and compare posterior means and SDs. Explain the differences.
3. Run the analysis for the larger data set for a long time, and from time to time, draw a density plot (in WinBUGS) or inspect the stats, for instance, after 10, 100, 1000, and 10 000 iterations: see how increasing length of Markov chains leads to improved estimates. Look at the MC error (in the WinBUGS output, hence you have to set `default = TRUE`) and at the values of the BGR statistic (=Rhat). Explain what you see.
4. Compare MC error and posterior SD and distinguish these things conceptually. See how they change in Markov chains of different lengths (as in Exercise 3).
5. *Derived quantities*: Estimate the coefficient of variation (CV = SD/mean) of body mass in that peregrine population. Report a point estimate (with SE) of that CV and also draw a picture of its posterior distribution.
6. *Swiss hare data*: Fit the model of the mean to *mean.density* and report the mean, the SE of the mean, and a 95% credible interval for mean hare density.

6

Key Components of (Generalized) Linear Models: Statistical Distributions and the Linear Predictor

Introduction to WinBUGS for Ecologists
DOI: 10.1016/B978-0-12-378605-0.00006-5

57

6.1 INTRODUCTION

All models in this book explain the variation in an observed response as being composed of a deterministic and a stochastic part. Thus, the concept of these models is this:

response = deterministic part + stochastic part

The deterministic part is also called the systematic part and the stochastic the random part of a model. It is the presence of the stochastic part that makes a model a *statistical*, rather than simply a mathematical model. For the description of the stochastic part of the model, we use a statistical distribution. In order to choose the "right" distribution, we need to know the typical sampling situation that leads to one rather than another distribution. Sampling situation means how the studied objects were chosen and how and what characteristic was measured. The quantity measured or otherwise assessed, the variation of which one wants to explain, is the response. In this chapter, we briefly review four of the most common statistical distributions that are used to capture the variability in a response: the normal, the uniform, the binomial, and the Poisson. These are virtually all the distributions that you need to know for this book. In addition, they are also those that underlie a vast range of statistical techniques that ecologists commonly employ.

Linear models are so called because they assume that the mean (i.e., the expected) response can be treated as the result of explanatory variables whose effects add together. Effects are additive regardless of whether the explanatory variables are continuous (covariates, regressors) or discrete (factors). To be able to specify how exactly we think the explanatory variables affect the response, we need to understand the so-called linear predictor of the model, the design matrix, and different parameterizations of a linear model. It is the design matrix along with the observed values of the covariates that makes up the deterministic part of a linear statistical model. They combine to form the linear predictor, i.e., the expected response, which is the value of the response that we would expect to observe in the absence of a stochastic component in the studied system.

In programs such as R or GenStat, the specification of the design matrix, and hence the linear predictor, just "happens" internally without us having to know how exactly this works. All we type is a formula to describe the linear model as in `response ~ effect1 + effect2`. In contrast and unfortunately (or actually, fortunately!), using WinBUGS requires us to know exactly what kind of linear model we want to fit because we have to describe this model at a very elementary level. Therefore, in this chapter, we review some of the basics of linear models. I expect them to be useful also for your general understanding of linear statistical models, whether analyzed in a Bayesian or frequentist mode of inference. We deal first with distributions (the stochastic part of a model) and then with the design matrix and the linear predictor (the deterministic part).

6.2 STOCHASTIC PART OF LINEAR MODELS: STATISTICAL DISTRIBUTIONS

Parametric statistical modeling means describing a caricature of the "machine" that plausibly could have produced the numbers we observe. The machine is nature, and *nature is stochastic*, i.e., nature is never fully predictable. An element of chance in the observed outcome of something, e.g., body mass, does not mean that the outcome has no reason for happening. There is always a reason; we simply don't know it. Chance just means that there are a few or many unrecognized factors that affect the outcome, but we either don't know or haven't measured them, or that we don't fully understand the relationship between these factors and an outcome. Things whose outcome is affected by chance and thus are only predictable up to a certain degree are called *random variables*. At some level, almost anything in nature is best thought of as a random variable.

However, random variables are seldom totally unpredictable; instead, the *combined effect* of all the unmeasured factors can often reasonably well be described by a mathematical abstraction: a probability distribution. A probability distribution assigns to each possible realization (value or event) of a random variable a probability of occurrence. To be a proper probability distribution, that description must be complete, i.e., all possible realizations must be described so that the sum of their probabilities of occurring is equal to 1.

The types of events vary greatly and may be binary like a coin flip (heads/tail), a survival event (dead/survived), or sex (male/female). They may be categorical like hair color (brown/blonde/red/white/gray), nationality (Swiss/French/Italian), or geographical location. (Note that a binary random variable is just a special case of a categorical one.) They may be counts, like the number of birds in a sampling quadrat, the number of people in a park or pub, or the number of boys in a school class (Note the slight difference in this last kind of count?). Or they can be measurements, like body mass, wing length, or diameter of a tree.

Probability distributions are themselves governed (described) by one or a few parameters, and their actual form depends on the particular values of these parameters. A probability distribution can be described using a formula, a table, or a picture. Statisticians usually prefer formulae and ecologists pictures. The advantage of a formula is that it is completely general, while a picture can only ever show how a distribution looks like for particular parameter values. Depending on these values, different distributions can yield very similar pictures or same distributions can look very different. Nevertheless, there are some typical features of many distributions, and it may be worthwhile to describe these, so you have a better chance of recognizing a distribution when you meet it.

Of the four distributions we encounter in this book, two are discrete and have non-negative values, i.e., they can only take on integer values from 0, 1, and upwards. They are the binomial distribution and the Poisson distribution. Then, we have two continuous distributions, i.e., that can take on any value, within measuring accuracy. One is the famous normal distribution, which is defined on the entire real line (from $-\infty$ to ∞), and the other is the uniform distribution, which is usually defined between its lower and upper limits.

Sometimes, people are astonished at how modelers know how to choose the right kind of distribution for describing their random variables. It is true that there is some arbitrary element involved and, often, there is more than a single distribution that could be used to describe the output of the machine that produced our observations. However, very frequently the circumstances of the data and how these were collected quite firmly point to one rather than another distribution as the most appropriate description of the number-generating machine and therefore also of the random variability in the output of that machine, the response.

Here, I describe circumstances in which each distribution would arise during sampling and then provide a pictorial example for each. I also give few lines of R code that are required to create random samples from each distribution and then plot them in a histogram for checking how selected parameter values influence the shape of a distribution. You can play around with different parameter values and so get a feel for how the distributions change as parameters are varied.

Among the four distributions (normal, uniform, Poisson, and binomial), one big divide is whether your response (i.e., the data) is discrete or continuous. Counts are discrete and point to the two latter distributions, while measurements are continuous and point to the two former distributions. In practice, there is some overlap. Since measurement accuracy is finite, every continuous random variable is recorded in a discrete way. However, this is usually of no consequence. On the other hand, under certain circumstances (e.g., large sample sizes, many observed unique values), the two discrete distributions can often be well approximated by a normal distribution. For instance, large counts in practice are often modeled as coming from a normal distribution.

What follows are brief vignettes of the four distributions we use throughout this book. Further features of these and many other distributions for selected parameter values can be studied in R (type `?rnorm`, `?runif`, `?rpois`, or `?rbinom` to find out more).

6.2.1 Normal Distribution

Sampling situation: Measurements are taken, which are affected by a large number of effects that act in an additive way.

Classical examples: They include body size and other linear measurements. In WinBUGS, they are also used to specify ignorance in priors when the precision is small (meaning the variance large).

Varieties: When effects are multiplicative instead of additive, we get the log-normal distribution. A log-normal random variable can be transformed into a normal one by log-transforming it.

Typical picture: It is the Gaussian bell curve, i.e., symmetrical, single hump, more or less long tails. In small samples, it can look remarkably irregular, e.g., skewed.

Mathematical description: It includes two parameters, the mean (location) and standard deviation (spread, average deviation from the mean) or, equivalently, the variance (squared standard deviation). In WinBUGS, spread is specified as precision = 1/variance.

Specification in WinBUGS:

```
x ~ dnorm(mean, tau) # note tau = 1/variance
```

R code to draw n *random number with specified parameter(s) and plot a histogram (see Fig. 6.1)*:

```
n <- 100000              # Sample size
mu <- mean <- 600        # Body mass of male peregrines
sd <- st.dev <- 30       # SD of body mass of male peregrines
```

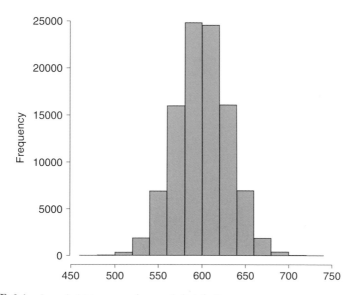

FIGURE 6.1   Sample histogram of normal distribution.

```
sample <- rnorm(n = n, mean = mu, sd = sd)
print(sample, dig = 4)
hist(sample, col = "grey")
```

Mean and standard deviation:

$$E(y) = \mu \qquad \text{mean}$$
$$\text{sd}(y) = \sigma \qquad \text{sd}$$

6.2.2 Continuous Uniform Distribution

Sampling situation: Measurements are taken, which are all equally likely to occur in a certain range of values.
Classical examples: In WinBUGS, this distribution is typically used to specify ignorance in a prior (as an alternative to a "flat" normal).
Varieties: The distribution can be discrete uniform, e.g., the result of rolling a die with the response being anything from 1 to 6.
Typical picture: It is a rectangle shape (with variation due to sampling variability).
Mathematical description: It includes two parameters, lower (a) and upper limits (b).
Specification in WinBUGS:

```
x ~ dunif(lower, upper)
```

R code to draw n *random number with specified parameter(s) and plot histogram (see Fig. 6.2)*:

```
n <- 100000            # Sample size
a <- lower.limit <- 0   #
b <- upper.limit <- 10  #

sample <- runif(n = n, min = a, max = b)
print(sample, dig = 3)
hist(sample, col = "grey")
```

Mean and standard deviation:

$$E(y) = (a + b)/2 \qquad \text{mean}$$
$$sd(y) = \sqrt{(b-a)^2/12} \qquad \text{sd}$$

6.2.3 Binomial Distribution: The "Coin-Flip Distribution"

Sampling situation: When N available things all have the same probability p of being in a certain state (e.g., being counted, or having a certain attribute, like being male or dead), then the number x that is actually counted in that sample, or has that attribute, is binomially distributed.

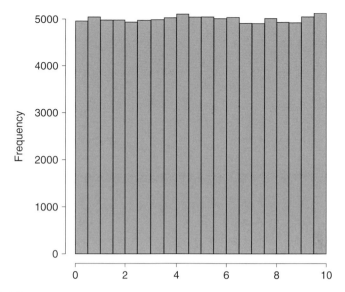

FIGURE 6.2   Sample histogram of uniform distribution.

Classical examples: Number of males in a clutch, school class, or herd of size N; number of times heads shows up among N flips of a coin; number of times you get a six among $N = 10$ rolls of a die; and the number of animals among those N present that you actually detect. *Varieties*: The Bernoulli distribution corresponds to a single coin flip and has only a single parameter, p. Actually, a binomial is the sum of N Bernoullis (or coin flips).

Typical picture: It varies a lot but strictly speaking always discrete. Normally, it is skewed, but skewness depends on the actual values of the parameters. The binomial distribution is symmetrical for $p = 0.5$.

Mathematical description: It includes one or two parameters, the probability of being chosen or having a certain trait (male, dead), often called success probability p, and the "binomial total" N, which is the sample or trial "size." N represents a ceiling to a binomial count; this is an important distinction to the similar Poisson distribution. Usually, N is observed and therefore not a parameter (but see Chapter 21). Aim of modeling is typically estimation and modeling of p (but sometimes also of N; see Chapter 21).

Important feature: The binomial comes with a "built-in" variance equal to $N * p * (1 - p)$, i.e., the variance is a function of the mean, which is $N * p$. See also Fig. 6.3.

Specification in WinBUGS:

```
x ~ dbin(p, N) # Note order of parameters !
```

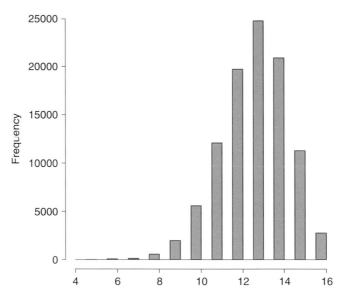

FIGURE 6.3   Sample histogram of binomial distribution with $N=16$ and $p=0.8$.

Bernoulli:

```
x ~ dbern(p)
```

R code to draw n *random number with specified parameter(s) and plot histogram (Fig. 6.3):*

```
n <- 100000          # Sample size
N <- 16              # Number of individuals that flip the coin
p <- 0.8             # Probability of being counted (seen), dead or a male
sample <- rbinom(n = n, size = N, prob = p)
print(sample, dig = 3)
hist(sample, col = "grey")
```

Extra comment: This picture can be thought to give the frequency with which we count 4, 5, 6, ..., 16 greenfinches in a monitoring plot that has a population of $N = 16$ individuals and where each greenfinch has a probability of being seen (= counted) of exactly $p = 0.8$. It is important, though not sufficiently widely recognized among ornithologists, to note that there is variation in bird counts even under constant conditions, i.e., when N and p are constant (Kéry and Schmidt, 2008).

Mean and standard deviation:

$$E(y) = N * p \qquad \text{mean}$$
$$sd(y) = \sqrt{N * p * (1-p)} \qquad \text{sd}$$

Mean and standard deviation for the Bernoulli are p and $\sqrt{p * (1-p)}$, respectively.

6.2.4 Poisson Distribution

Sampling situation: When things (e.g., birds, cars, or erythrocytes) are randomly distributed in one or two (or more) dimensions and we randomly place a "counting window" along that dimension or in that space and record the number of things, then that number is Poisson distributed.

Classical examples: Number of passing cars during 10-min counts at a street corner; number of birds that fly by you at a migration site, annual number of Prussian soldiers kicked to death by a horse; number of car accidents per day, month, or year, and number of birds or hares per sample quadrat.

Varieties: None really. But the Poisson is an approximation to the binomial when N is large and p is small and can itself be approximated by the normal when the average count, λ, is large, e.g., greater than 10. The negative binomial distribution is an overdispersed version of the Poisson and can be derived by assuming that the Poisson mean, λ, is itself a random variable with another distribution, the gamma. Therefore, it is also referred to as a Poisson–gamma mixture.

Typical picture: It varies a lot but strictly speaking always discrete. It is skewed normally, but skewness depends on the value of lambda.

Mathematical description: It includes a single parameter called λ, which is equal to the mean (= expectation, average count, intensity), as well as the variance (i.e., variance = mean). That is, as for the binomial distribution, the variance is not a free parameter but is a function of the mean.

Important feature: As for the binomial, values from a Poisson distribution are non-negative integers that come with a built-in variance.

Specification in WinBUGS:

```
x ~ dpois(lambda)
```

R code to draw n *random number with specified parameter(s) and plot histogram (see Fig. 6.4):*

```
n <- 100000        # Sample size
lambda <- 5        # Average # individuals per sample, density
sample <- rpois(n = n, lambda = lambda)
print(sample, dig = 3)

par(mfrow = c(2,1))
hist(sample, col = "grey", main = "Default histogram")
plot(table(sample), main = "A better graph", lwd = 3, ylab = "Frequency")
```

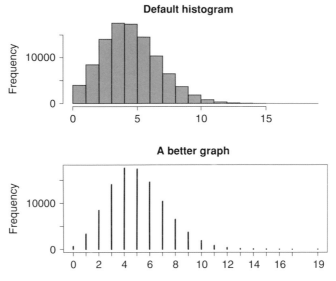

FIGURE 6.4   Sample histogram of Poisson distribution. The lower graph is better since it shows the discrete nature of a Poisson random variable.

Mean and standard deviation:

$$E(y) = \lambda \qquad \text{mean}$$
$$sd(y) = \sqrt{\lambda} \qquad \text{sd}$$

6.3 DETERMINISTIC PART OF LINEAR MODELS: LINEAR PREDICTOR AND DESIGN MATRICES

Linear models describe the expected response as a linear combination of the effects of discrete or continuous explanatory variables. That is, they directly specify the relationship between the response and one or more explanatory variables. I note, in passing, that it is this linear (additive) relationship that makes a statistical model linear, not that a picture of it is a straight line. Most statistically linear models describe relationships that are represented by curvilinear graphs of some kind. Thus, one must differentiate between linear statistical models and linear models that also imply straight-line relationships.

Most ecologists have become accustomed to using a simple formula language when specifying linear models in statistical software. For instance, to specify a two-way analysis of variance (ANOVA) with interaction in R, we just type y ~ A*B as an argument of the functions lm() or glm(). Many of us don't know exactly what typing y ~ A*B causes R to do internally, and this may not be fatal. However, to fit a linear model in

WinBUGS, you must know exactly what this and other model descriptions mean and how they can be adapted to match your hypotheses! Therefore, I next present a short guide to the so-called design matrix of a linear or generalized linear model (GLM). We will see how the same linear model can be described in different ways; the different forms of the associated design matrices are called *parameterizations* of a model.

The material in the rest of this chapter may be seen by some as a nuisance. It may look difficult at first and indeed may not be totally necessary when fitting linear models using one of the widely known stats packages. For better or worse, in order to use WinBUGS, you must understand this material. However, understanding design matrices will greatly increase your grasp of statistical models in general. In particular, you will greatly benefit from this understanding when you use other software to conduct more specialist analyses that ecologists typically conduct. For instance, you need to understand the design matrix when you use program MARK (White and Burnham, 1999) to fit any of a very large range of capture–recapture types of models (for instance, most of those described by Williams et al., 2002). Thus, time invested to understand this material is time well spent for most ecologists!

For each element of the response vector, the design matrix n index indicates which effect is present for categorical (= discrete) explanatory variables and what "amount" of an effect is present in the case of continuous explanatory variables. The design matrix contains as many columns as the fitted model has parameters, and when matrix-multiplied with the parameter vector, it yields the linear predictor, another vector. The linear predictor contains the expected value of the response (on the link scale), given the values of all explanatory variables in the model. Expected value means the response that would be observed when all random variation is averaged out. For a fuller understanding of some of this GLM jargon, you may need to jump to Chapter 13 (and back again). However, this is not required for an understanding of this chapter.

In the remainder of this chapter, we look at a progression of typical linear models (e.g., t-test, simple linear regression, one-way ANOVA, and analysis of covariance (ANCOVA)). We see how to specify them in R and then find out what R does internally when we do this. We look at how to write these models algebraically and what system of equations they imply. We see different ways of writing what is essentially the same model (i.e., different parameterizations of a model) and how these affect the interpretation of the model parameters. Only understanding all the above allows us to specify these linear models in WinBUGS (though, notably, none of this has anything to do with Bayesian statistics).

For most nonstatisticians, it is usually much easier to understand something with the aid of a numerical example. So to study the design matrices for these linear models, we introduce a toy data set consisting of just six

TABLE 6.1   Our Toy Data Set for Six Snakes

mass	pop	region	hab	svl
6	1	1	1	40
8	1	1	2	45
5	2	1	3	39
7	2	1	1	50
9	3	2	2	52
11	3	2	3	57

data points (Table 6.1). Let's imagine that we measured body mass (mass, in units of 10 g; a continuous response variable) of six snakes in three populations (pop), two regions (region), and three habitat types (hab). These three discrete explanatory variables are factors with 3, 2, and 3 levels, respectively. We also measured a continuous explanatory variable for each snake, snout–vent length (svl).

Here is the R code for setting up these variables:

```
mass <- c(6, 8, 5, 7, 9, 11)
pop <- factor(c(1,1,2,2,3,3))
region <- factor(c(1,1,1,1,2,2))
hab <- factor(c(1,2,3,1,2,3))
svl <- c(40, 45, 39, 50, 52, 57)
```

We use factor() to tell R that the numbers in this variable are just names that lack any quantitative meaning. Next, we use the R functions lm() or glm() to specify different linear models for these data and inspect what this exactly means in terms of the design matrix.

6.3.1 The Model of the Mean

What if we just wanted to fit a common mean to the mass of all six snakes? That is, we want to fit a linear model with an intercept only (see Chapters 4 and 5). In R, this is done simply by issuing the command:

```
lm(mass ~ 1)
```

The 1 implies a covariate with a single value of one for every snake. One way to write this model algebraically is this:

$$\text{mass}_i = \mu + \varepsilon_i$$

That is, we imagine that the mass measured for snake i is composed of an overall mean, μ, plus an individual deviation from that mean called ε_i. Of course, the latter is the residual for snake i. If we want to estimate the mean within a linear statistical model as above, then we also need an assumption about these residuals, for instance $\varepsilon_i \sim \text{Normal}(0, \sigma^2)$. That is, we assume that the residuals are normally distributed around μ with a variance of σ^2.

Let's look at the design matrix of that model using the function `model.matrix()`. We see that it consists just of a vector of ones and that this vector is termed the intercept by R.

```
model.matrix(mass ~ 1)
```

In the rest of this book, we make extensive use of this very useful R function, `model.matrix()`. Next, we consider in more detail some less trivial linear statistical models.

6.3.2 t-Test

When we are interested in the effect of a single, binary explanatory variable such as `region` on a continuous response such as `mass`, we can conduct a t-test (see Chapter 7). Here is the specification of a t-test as a linear model in R:

```
lm(mass ~ region)
```

Some might wonder now: "But where is the p-value?" However, the most important thing in the linear model underlying a t-test is not the p-value but the vector of parameter estimates. R is a respectable statistics program and thus presents the most important things first. But of course, there are several ways the p-value might be obtained, for instance, by `summary(lm(mass ~ region))`.

In the methods section of a research paper, you would not describe your model with the R description above; rather, you could describe it algebraically. Of course, for such a simple model as that underlying the t-test, you would simply mention that you used a t-test, but for the sake of getting some mental exercise, let's do it now in a more general way. One way of describing the linear model that relates snake mass to region is with the following statement:

$$\text{mass}_i = \alpha + \beta * \text{region}_i + \varepsilon_i$$

This means that the mass of snake i is made up of the sum of three components: a constant α, the product of another constant (β) with the value of the indicator for the region in which snake i was caught, and a third term ε_i that is specific to each snake i. The last term, ε_i, is the residual, and to make this mathematical model a statistical model, we need some assumption about how these residuals vary among individuals. One of the simplest assumptions about these unexplained, snake-specific contributions to mass is that they follow a normal distribution centered on zero and with a spread (standard deviation or variance) that we are going to estimate. We can write this as follows:

$$\varepsilon_i \sim \text{Normal}(0, \sigma^2)$$

This way of writing the linear model underlying a t-test immediately clarifies a widespread confusion about what exactly needs to be normally distributed in normal linear models: it is the residuals and not the raw response. Thus, if you are concerned about the normality or not of a response variable, you must first fit a linear model and only then inspect the residual distribution.

A second and equivalent way to write the linear model for the t-test algebraically is this:

$$\text{mass}_i \sim \text{Normal}(\alpha + \beta * \text{region}_i, \sigma^2)$$

This way to write down the model resembles more the way in which we specify it in WinBUGS, as we will see. It clarifies that the mean response is not the same for each snake. Moreover, in this way to write the model, $\alpha + \beta * \text{region}_i$ represents the deterministic part of the model and $\text{Normal}(..., \sigma^2)$ its stochastic part.

Finally, we might also write this model like this: $\text{mass}_i = \mu_i + \varepsilon_i$, which again says that the mass of snake i is the sum of a systematic effect, embodied by the mean or expected mass μ_i, and a random effect ε_i. The expected mass of snake i, μ_i, consists of $\alpha + \beta * \text{region}_i$. This is called the linear predictor of the model, and the residuals are assumed to follow a normal distribution centered on the value of the linear predictor, as before.

Regardless of how we describe the model, we must know what the two-level variable named region actually looks like in the analysis of the linear model underlying the t-test. When we define region to be a factor and then fit its effect, what R does internally is to convert it into a design matrix with two columns, one containing only ones (corresponding to the intercept) and the other being a single indicator or dummy variable. That is a variable containing just ones and zeroes that indicates which snake was caught in the region indicated in R's name for that column in the design matrix.

We can look at this using the function model.matrix():

```
model.matrix(~region)
> model.matrix(~region)
  (Intercept)    region2
1           1          0
2           1          0
3           1          0
```

4	1	0
5	1	1
6	1	1

[...]

(This is what the R default *treatment contrasts* yield; type ?model. matrix to find out what other contrasts are possible in R.) Therefore, the indicator variable named region2 contains a one for those snakes in region 2.

Fitting the model underlying the t-test to our six data points with region defined in this way implies a system of equations, and it is very instructive to write it down:

$$6 = \alpha * 1 + \beta * 0 + \varepsilon_1$$
$$8 = \alpha * 1 + \beta * 0 + \varepsilon_2$$
$$5 = \alpha * 1 + \beta * 0 + \varepsilon_3$$
$$7 = \alpha * 1 + \beta * 0 + \varepsilon_4$$
$$9 = \alpha * 1 + \beta * 1 + \varepsilon_5$$
$$11 = \alpha * 1 + \beta * 1 + \varepsilon_6$$

Here, one sees why the intercept is often represented by a 1: it is always present and identical for all the values of the response variable. To get a solution for this system of equations, i.e., to obtain values for the unknowns α and β that are "good" in some way, we need to define some criterion for dealing with the residuals ε_i. Usually, in this system of equations, the unknowns α and β are chosen such that the sum of the squared residuals is minimal. This is called the least-squares method, and for normal GLMs (again, see Chapter 13), the resulting parameter estimates for α and β are equivalent to those obtained using the more general maximum likelihood method.

A more concise way of writing the same system of equations is using vectors and matrices:

$$\begin{pmatrix} 6 \\ 8 \\ 5 \\ 7 \\ 9 \\ 11 \end{pmatrix} = \begin{pmatrix} 1 & 0 \\ 1 & 0 \\ 1 & 0 \\ 1 & 0 \\ 1 & 1 \\ 1 & 1 \end{pmatrix} * \begin{pmatrix} \alpha \\ \beta \end{pmatrix} + \begin{pmatrix} \varepsilon_1 \\ \varepsilon_2 \\ \varepsilon_3 \\ \varepsilon_4 \\ \varepsilon_5 \\ \varepsilon_6 \end{pmatrix}$$

We see again that the response is made up of the value of the linear predictor plus the vector of residuals. The linear predictor consists of the product of the *design matrix* (also called model matrix or X matrix) and the *parameter vector*. In a sense, the design matrix contains the "weights" with which α and β enter the linear predictor for each snake i. The linear predictor, i.e., the result of that matrix multiplication, is another vector containing the expected value of the response, given the covariates, for each snake *i*.

What is the interpretation of these parameters? This is a crucial topic and one that depends entirely on the way in which the design matrix is specified for a model. It is clear from looking at the extended version of the system of equations above (or the matrix version as well) that α must represent the expected mass of a snake in region 1 and β is the difference in the expected mass of a snake in region 2 compared with that of a snake in region 1. Thus, region 1 serves as a baseline or reference level and in the model becomes represented by the intercept parameter, and the *effect* of region 2 is parameterized as a difference from that baseline. Let's look again at an analysis of this model parameterization in R:

```
lm(mass ~ region)
> lm(mass ~ region)

Call:
lm(formula = mass ~ region)

Coefficients:
(Intercept)    region2
        6.5        3.5
```

We recognize the value of the intercept as the mean mass of snakes in region 1 and the parameter called `region2` as the difference between the mass in region 2 and that in region 1, i.e., the effect of region 2, or here the effect of region. Hence, we may call this an *effects parameterization* of the t-test model. Actually, in the output, R labels the parameter for region 2 by the name of its column in the design matrix and that may be slightly misleading.

Importantly, this is not the only way in which a linear model representing a t-test for a difference between the two regions might be specified. Another way how to set up the equations, which is equivalent to the one above and is called a *reparameterization* of that model, is this:

$$\begin{pmatrix} 6 \\ 8 \\ 5 \\ 7 \\ 9 \\ 11 \end{pmatrix} = \begin{pmatrix} 1 & 0 \\ 1 & 0 \\ 1 & 0 \\ 1 & 0 \\ 0 & 1 \\ 0 & 1 \end{pmatrix} * \begin{pmatrix} \alpha \\ \beta \end{pmatrix} + \begin{pmatrix} \varepsilon_1 \\ \varepsilon_2 \\ \varepsilon_3 \\ \varepsilon_4 \\ \varepsilon_5 \\ \varepsilon_6 \end{pmatrix}$$

What do the parameters α and β mean now? The interpretation of α has not changed; it is the expected mass of a snake in region 1. However, the interpretation of parameter β is different, since now it is the expected body mass of a snake in region 2. Thus, in this parameterization of the linear model underlying the t-test, the parameters directly represent the group means, and we may call this a *means parameterization* of the model. This model can be fitted in R by removing the intercept:

```
model.matrix(~region-1)
> model.matrix(~region-1)
   region1  region2
1       1        0
2       1        0
3       1        0
4       1        0
5       0        1
6       0        1
[ ... ]

lm(mass~region-1)
> lm(mass~region-1)

Call:
lm(formula = mass ~ region - 1)

Coefficients:
region1    region2
    6.5       10.0
```

Why would we want to use these different parameterizations of the same model? The answer is that they serve different aims: the effects parameterization is more useful for testing for a difference between the means in the two regions; this is equivalent to testing whether the effect of region 2, i.e., parameter b, is equal to zero. In contrast, for a summary of the analysis, we might prefer the means parameterization and directly report the estimated expected mass of snakes for each region. However, the two models are equivalent; for instance, the sum of the two effects in the former parameterization is equal to the value of the second parameter in the latter parameterization, i.e., $6.5 + 3.5 = 10$.

6.3.3 Simple Linear Regression

To examine the relationship between a continuous response (mass) and a continuous explanatory variable such as svl, we would specify a simple linear regression in R (see also Chapter 8):

```
lm(mass ~ svl)
```

This model is written algebraically in the same way as that underlying the t-test:

$$\text{mass}_i = \alpha + \beta * \text{svl}_i + \varepsilon_i$$

and

$$\varepsilon_i \sim \text{Normal}(0, \sigma^2)$$

Or like this:

$$\text{mass}_i \sim \text{Normal}(\alpha + \beta * \text{svl}_i, \sigma^2)$$

The only difference to the t-test lies in the contents of the explanatory variable, svl, which may contain any real number rather than just the two values of 0 and 1. Here, svl contains the lengths for each of the six snakes. We inspect the design matrix that R builds on issuing the above call to lm():

```
model.matrix(~svl)
> model.matrix(~svl)
  (Intercept)   svl
1           1    40
2           1    45
3           1    39
4           1    50
5           1    52
6           1    57
[]
```

Thus, fitting this simple linear regression model to our six data points implies solving the following system of equations:

$$6 = \alpha + \beta * 40 + \varepsilon_1$$
$$8 = \alpha + \beta * 45 + \varepsilon_2$$
$$5 = \alpha + \beta * 39 + \varepsilon_3$$
$$7 = \alpha + \beta * 50 + \varepsilon_4$$
$$9 = \alpha + \beta * 52 + \varepsilon_5$$
$$11 = \alpha + \beta * 57 + \varepsilon_6$$

Again, a more concise way of writing the same equations is by using vectors and a matrix:

$$\begin{pmatrix} 6 \\ 8 \\ 5 \\ 7 \\ 9 \\ 11 \end{pmatrix} = \begin{pmatrix} 1 & 40 \\ 1 & 45 \\ 1 & 39 \\ 1 & 50 \\ 1 & 52 \\ 1 & 57 \end{pmatrix} * \begin{pmatrix} \alpha \\ \beta \end{pmatrix} + \begin{pmatrix} \varepsilon_1 \\ \varepsilon_2 \\ \varepsilon_3 \\ \varepsilon_4 \\ \varepsilon_5 \\ \varepsilon_6 \end{pmatrix}$$

The design matrix contains an intercept and another column with the values of the covariate svl. The interpretation of the parameters α and β is thus that of a baseline, representing the expected value of the response (mass) at a covariate value of svl = 0, and a difference or an effect, representing the change in mass for each unit change in svl. Equivalently, this effect is the slope of the regression of mass on svl or the effect of svl.

Let's look at an analysis of this model in R:

```
lm(mass~svl)
> lm(mass~svl)

Call:
lm(formula = mass ~ svl)

Coefficients:
(Intercept)        svl
    -5.5588     0.2804
```

The intercept is biologically nonsense; it says that a snake of zero length weighs −5.6 mass units! This illustrates that a linear model can only ever be a useful characterization of a biological relationship over a restricted range of the explanatory variables. To give the intercept a more sensible interpretation, the model could be reparameterized by transforming svl to svl-mean(svl). Fitting this centered version of svl will cause the intercept to become the expected mass of a snake at the average of the observed size distribution.

Again, this is not the only way to specify a linear regression of mass on svl, and we could again remove the intercept as in the t-test. However, removing the intercept is not just a reparameterization of the same model; instead, it changes the model and forces the regression line to go through the origin. The design matrix then contains just the values of the covariate svl:

```
model.matrix(~svl -1)
> model.matrix(~svl-1)
  svl
1   40
2   45
3   39
4   50
5   52
6   57
[ ... ]

lm(mass~svl-1)
> lm(mass~svl-1)

Call:
lm(formula = mass ~ svl - 1)

Coefficients:
   svl
0.1647
```

We see that the estimated slope is less than before, which makes sense, since previously the estimate of the intercept was negative and now it is forced to be zero. In most instances, the no-intercept model is not very sensible, and we should have strong reasons for forcing the intercept of a linear regression to be zero. Also, the lower limit of the observed range of the covariate should be close to, or include, zero.

6.3.4 One-Way Analysis of Variance

To examine the relationship between mass and pop, a factor with three levels, we would specify a one-way ANOVA in R:

```
lm(mass ~ pop)
```

There are different equivalent ways in which to write this model algebraically; see also Chapter 9. Here, we focus on the effects and the means parameterizations and show how their design matrices look. For the mass of individual i in population j, one way to write the model is like this:

$$\text{mass}_i = \alpha + \beta_{j(i)} * \text{pop}_i + \varepsilon_i$$

and

$$\varepsilon_i \sim \text{Normal}(0, \sigma^2)$$

Another way to specify the same model is this:

$$\text{mass}_i \sim \text{Normal}(\alpha + \beta_{j(i)} * \text{pop}_i, \sigma^2)$$

This parameterization means to set up the design matrix in an *effects* format, i.e., with an intercept α, representing the mean for population 1, plus indicators for every population except the first one, representing the differences between the means in these populations and the means in the reference population. (The choice of reference is arbitrary and has no influence on inference.). For n populations, there are $n-1$ β_j parameters, which are indexed by factor level j.

An algebraic description of the *means* parameterization of the one-way ANOVA would be this:

$$y_i = \alpha_{j(i)} * \text{pop}_i + \varepsilon_i$$

and

$$\varepsilon_i \sim \text{Normal}(0, \sigma^2)$$

Here, the design matrix is set up to indicate each individual population, and the parameters α_j represent the mean body mass of snakes in each population j.

We look at the two design matrices associated with the two parameterizations.

First, the effects parameterization, which is the default in R:

```
model.matrix(~pop)
> model.matrix(~pop)
  (Intercept)  pop2  pop3
1           1     0     0
2           1     0     0
3           1     1     0
4           1     1     0
5           1     0     1
6           1     0     1
[ ... ]
```

Second, the means parameterization, which is specified in R by removing the intercept:

```
model.matrix(~pop-1)
> model.matrix(~pop-1)
  pop1  pop2  pop3
1    1     0     0
2    1     0     0
3    0     1     0
4    0     1     0
5    0     0     1
6    0     0     1
[ ... ]
```

Again, the interpretation of the first parameter of the model is the same for both parameterizations: it is the mean body mass in population 1. However, while in the effects parameterization, the parameters 2 and 3 correspond to differences in the means (above), and they represent the actual expected body mass for snakes in each of the three populations in the means parameterization (below). Also note how the first parameter changes names when going from the former to the latter parameterization of the one-way ANOVA.

Here is the matrix–vector description of the effects ANOVA model applied to our toy snake data set:

$$
\begin{pmatrix} 6 \\ 8 \\ 5 \\ 7 \\ 9 \\ 11 \end{pmatrix} = \begin{pmatrix} 1 & 0 & 0 \\ 1 & 0 & 0 \\ 1 & 1 & 0 \\ 1 & 1 & 0 \\ 1 & 0 & 1 \\ 1 & 0 & 1 \end{pmatrix} * \begin{pmatrix} \alpha \\ \beta_2 \\ \beta_3 \end{pmatrix} + \begin{pmatrix} \varepsilon_1 \\ \varepsilon_2 \\ \varepsilon_3 \\ \varepsilon_4 \\ \varepsilon_5 \\ \varepsilon_6 \end{pmatrix}
$$

And here is the means parameterization:

$$
\begin{pmatrix} 6 \\ 8 \\ 5 \\ 7 \\ 9 \\ 11 \end{pmatrix} = \begin{pmatrix} 1 & 0 & 0 \\ 1 & 0 & 0 \\ 0 & 1 & 0 \\ 0 & 1 & 0 \\ 0 & 0 & 1 \\ 0 & 0 & 1 \end{pmatrix} * \begin{pmatrix} \alpha_1 \\ \alpha_2 \\ \alpha_3 \end{pmatrix} + \begin{pmatrix} \varepsilon_1 \\ \varepsilon_2 \\ \varepsilon_3 \\ \varepsilon_4 \\ \varepsilon_5 \\ \varepsilon_6 \end{pmatrix}
$$

Again, we see clearly that the interpretation of the first parameter in the model is not affected by the parameterization chosen, but those for the second and third are.

Let's fit the two versions of the ANOVA in R, first the effects and then the means model:

```
lm(mass~pop)                    # Effects parameterization (R default)
> lm(mass~pop)

Call:
lm(formula = mass ~ pop)

Coefficients:
(Intercept)          pop2        pop3
          7            -1           3
lm(mass~pop-1)                  # Means parameterization
> lm(mass~pop-1)

Call:
lm(formula = mass ~ pop - 1)

Coefficients:
pop1  pop2  pop3
   7     6    10
```

Each parameterization is better suited to a different aim: the effects model is better for testing for differences and the means model is better for presentation.

6.3.5 Two-Way Analysis of Variance

A two-way or two-factor ANOVA serves to examine the relationship between a continuous response, such as mass, and two discrete explanatory variables, such as population (region) and habitat (hab), in our example (see also Chapter 10). Importantly, there are two different ways in which to combine the effects of two explanatory variables: additive (also called main effects) and multiplicative (also called interaction effects). In addition, there is the possibility to specify these models using an effects parameterization or a means parameterization. We consider each one in turn.

To specify a main-effects ANOVA with `region` and `hab` in R, we would write

```
lm(mass ~ region + hab)
> lm(mass ~ region + hab)

Call:
lm(formula = mass ~ region + hab)

Coefficients:
(Intercept)      region2      hab2      hab3
       6.50         3.50      0.25     -0.25
```

To avoid overparameterization (i.e., trying to estimate more parameters than the available data allow us to do), we need to arbitrarily set to zero the effects of one level for each factor. The effects of the remaining levels then get the interpretation of differences relative to the base level. It does not matter which level is used as a baseline or reference, but often stats programs use the first or the last level of each factor. R sets the effects of the first level to zero.

In our snake toy data set, for the mass of individual i in region j and habitat k, we can write this model as follows:

$$\text{mass}_i = \alpha + \beta_{j(i)} * \text{region}_i + \delta_{k(i)} * \text{hab}_i + \varepsilon_i$$

and

$$\varepsilon_i \sim \text{Normal}(0, \sigma^2)$$

Here, α is the expected mass of a snake in habitat 1 and region 1. There is only one parameter β_j, so the subscript could as well be dropped. It specifies the difference in the expected mass between snakes in region 2 and snakes in region 1. We need two parameters δ_k to specify the differences in the expected mass for snakes in habitats 2 and 3, respectively, relative to those in habitat 1.

We look at the design matrix

```
model.matrix(~region + hab)

> model.matrix(~region + hab)
  (Intercept)  region2  hab2  hab3
1           1        0     0     0
2           1        0     1     0
3           1        0     0     1
4           1        0     0     0
5           1        1     1     0
6           1        1     0     1
[ ... ]
```

Hence, the implied system of equations that the software solves for us is this:

$$
\begin{pmatrix} 6 \\ 8 \\ 5 \\ 7 \\ 9 \\ 11 \end{pmatrix}
=
\begin{pmatrix}
1 & 0 & 0 & 0 \\
1 & 0 & 1 & 0 \\
1 & 0 & 0 & 1 \\
1 & 0 & 0 & 0 \\
1 & 1 & 1 & 0 \\
1 & 1 & 0 & 1
\end{pmatrix}
*
\begin{pmatrix} \alpha \\ \beta_2 \\ \delta_2 \\ \delta_3 \end{pmatrix}
+
\begin{pmatrix} \varepsilon_1 \\ \varepsilon_2 \\ \varepsilon_3 \\ \varepsilon_4 \\ \varepsilon_5 \\ \varepsilon_6 \end{pmatrix}
$$

Interestingly, there is no way to specify the main-effects model in a means parameterization.

Next, consider the model with interactive effects, which lets the effect of one factor level depend on the level of the other factor. The default effects parameterization is written like this in R:

```
lm(mass ~ region * hab)
> lm(mass ~ region * hab)

Call:
lm(formula = mass ~ region * hab)

Coefficients:
(Intercept)   region2   hab2    hab3    region2:hab2   region2:hab3
6.5           6.0       1.5     -1.5    -5.0           NA
```

We see that one parameter is not estimable. The reason for this is that in the 2-by-3 table of effects of region crossed with habitat, we lack snake observations for habitat 1 in region 2.

Algebraically, we assume that the mass of individual i in region j and habitat k can be broken down as follows:

$$
\text{mass}_i = \alpha + \beta_{j(i)} * \text{region}_i + \delta_{k(i)} * \text{hab}_i + \gamma_{jk(i)} * \text{region}_i * \text{hab}_i + \varepsilon_i
$$

with

$$
\varepsilon_i \sim \text{Normal}(0, \sigma^2).
$$

In this equation, the meanings of parameters α, β_j, and δ_k remain as before, i.e., they specify the main effects of the levels for the habitat and region factors. The new coefficients, γ_{jk}, of which there are two, specify the *interaction effects* between these two factors.

Here is the design matrix for this model:

```
model.matrix(~region * hab)

> model.matrix(~region * hab)
  (Intercept)   region2   hab2   hab3   region2:hab2   region2:hab3
1           1         0      0      0              0              0
2           1         0      0      1              0              0
3           1         0      0      1              0              0
4           1         0      0      0              0              0
```

| 5 | 1 | 1 | 1 | 0 | 1 | 0 |
| 6 | 1 | 1 | 0 | 1 | 0 | 1 |

[...]

And here is, therefore, the system of equations that needs to be solved to get the parameter estimates for this model:

$$\begin{pmatrix} 6 \\ 8 \\ 5 \\ 7 \\ 9 \\ 11 \end{pmatrix} = \begin{pmatrix} 1 & 0 & 0 & 0 & 0 & 0 \\ 1 & 0 & 1 & 0 & 0 & 0 \\ 1 & 0 & 0 & 1 & 0 & 0 \\ 1 & 0 & 0 & 0 & 0 & 0 \\ 1 & 1 & 1 & 0 & 1 & 0 \\ 1 & 1 & 0 & 1 & 0 & 1 \end{pmatrix} * \begin{pmatrix} \alpha \\ \beta_2 \\ \delta_2 \\ \delta_3 \\ \gamma_{22} \\ \gamma_{23} \end{pmatrix} + \begin{pmatrix} \varepsilon_1 \\ \varepsilon_2 \\ \varepsilon_3 \\ \varepsilon_4 \\ \varepsilon_5 \\ \varepsilon_6 \end{pmatrix}$$

Finally, the means parameterization of the interaction model:

```
lm(mass ~ region * hab-1-region-hab)
> lm(mass ~ region * hab-1-region-hab)

Call:
lm(formula = mass ~ region * hab - 1 - region - hab)

Coefficients:
reg1:hab1  reg2:hab1  reg1:hab2  reg2:hab2  reg1:hab3  reg2:hab3
6.5        NA         8.0        9.0        5.0        11.0
```

(I slightly edited the output so it fits a single line.)

We see again that in the interactive model, we have no information to estimate the expected body mass of snakes in habitat of type 1 in region 2, since no snakes were examined for that combination of factor levels (called a *cell* in the cross-classification of the two factors). In the additive model, the information for that cell in the table comes from the other cells in the table, but in the interactive model, each cell mean is estimated independently.

This model can be written algebraically like this:

$$\text{mass}_i = \alpha_{jk(i)} * \text{region}_i * \text{hab}_i + \varepsilon_i$$

and

$$\varepsilon_i \sim \text{Normal}(0, \sigma^2)$$

There are six elements in the vector α_{jk}, corresponding to the six ways in which the levels of the two factors region and habitat can be combined. Also note that $region_i * hab_i$ implies six columns in the design matrix. Now look at the design matrix (again, slightly edited for an improved presentation):

```
model.matrix(~ region * hab-1-region-hab)
> model.matrix(~ region * hab-1-region-hab)
  reg1:hab1  reg2:hab1  reg1:hab2  reg2:hab2  reg1:hab3  reg2:hab3
1 1    1        0          0          0          0          0
2 0    0        0          1          0          0          0
```

3	0	0	0	0	1	0
4	1	0	0	0	0	0
5	0	0	0	1	0	0
6	0	0	0	0	0	1

[...]

And the system of equations:

$$\begin{pmatrix} 6 \\ 8 \\ 5 \\ 7 \\ 9 \\ 11 \end{pmatrix} = \begin{pmatrix} 1 & 0 & 0 & 0 & 0 & 0 \\ 0 & 0 & 1 & 0 & 0 & 0 \\ 0 & 0 & 0 & 0 & 1 & 0 \\ 1 & 0 & 0 & 0 & 0 & 0 \\ 0 & 0 & 0 & 1 & 0 & 0 \\ 0 & 0 & 0 & 0 & 0 & 1 \end{pmatrix} * \begin{pmatrix} \alpha_{11} \\ \alpha_{21} \\ \alpha_{12} \\ \alpha_{22} \\ \alpha_{13} \\ \alpha_{23} \end{pmatrix} + \begin{pmatrix} \varepsilon_1 \\ \varepsilon_2 \\ \varepsilon_3 \\ \varepsilon_4 \\ \varepsilon_5 \\ \varepsilon_6 \end{pmatrix}$$

We see clearly the lack of information about effect α_{21}, which is the expected mass of a snake in region 2 and habitat 1, represented by the second column in the design matrix containing all zeroes.

6.3.6 Analysis of Covariance

When we are interested in the effects on mass of both a factor (= discrete explanatory variable, e.g., pop) and a continuous covariate like svl, we could specify an ANCOVA model (see also Chapter 11). There are two ways in which we might like to specify that model. First, we may think that the relationship between mass and svl is the same in all populations, or worded in another way, that the mass differences among populations do not depend on the length of the snake. In statistical terms, this would be represented by a *main-effects model*. Second if we admitted that the mass–length relationship might differ among populations or that the differences in mass among populations might depend on the length of a snake, we would fit an *interaction-effects model*. (Note that here "effects" has a slightly different meaning from that when used as effects parameterization vs. means parameterization.) To specify these two versions of an ANCOVA in R, we write this:

```
lm(mass ~ pop + svl)          # Additive model
lm(mass ~ pop * svl)          # Interactive model
lm(mass ~ pop + svl + pop:svl) # Same, R's way of specifying the interaction term
```

Here's the additive model algebraically in the effects parameterization:

$$\text{mass}_i = \alpha + \beta_{j(i)} * \text{pop}_i + \delta * \text{svl}_i + \varepsilon_i,$$

with

$$\varepsilon_i \sim \text{Normal}(0, \sigma^2)$$

This model says that the mass of snake i in population j is made up of a constant α plus the effects β_j when in populations 2 and 3 plus a constant δ times svl plus the residual. In the effects parameterization, α is the intercept for population 1 and vector β_j has two elements, one for population 2 and another one for population 3, being the differences of the intercepts in these populations and the intercept in population 1. Finally, δ is the common slope of the mass–length relationship in all three populations.

The means parameterization of the same model is this:

$$\text{mass}_i = \alpha_{j(i)} * \text{pop}_i + \delta * \text{svl}_i + \varepsilon_i,$$

with

$$\varepsilon_i \sim \text{Normal}(0, \sigma^2)$$

All that changes is that now the vector α_j has three elements representing the intercepts in each population.

The effects parameterization of the interaction-effects model is written like this:

$$\text{mass}_i = \alpha + \beta_{j(i)} * \text{pop}_i + \delta * \text{svl}_i + \gamma_{j(i)} * \text{svl}_i * \text{pop}_i + \varepsilon_i$$

and

$$\varepsilon_i \sim \text{Normal}(0, \sigma^2)$$

In this model, α is the intercept for population 1 and vector β_j has two elements, representing the difference in the intercept between population 1 and populations 2 and 3, respectively. Parameter δ is the slope of the mass–length relationship in the first population. Vector γ_j has two elements corresponding to the difference in the slope between population 1 and populations 2 and 3, respectively.

The means parameterization of the same model is probably easier to understand and is written like this:

$$\text{mass}_i = \alpha_{j(i)} * \text{pop}_i + \delta_{j(i)} * \text{svl}_i + \varepsilon_i$$

and

$$\varepsilon_i \sim \text{Normal}(0, \sigma^2)$$

The only change relative to the main-effects model is that we added a subscript j to the effect δ of svl, meaning that we now estimate three slopes δ, one for each population, instead of a single one that is common to snakes from all three populations.

Let's now look at the design matrices of both the effects and the mean parameterizations of both the main- and interaction-effects models. Here is the design matrix of the main-effects ANCOVA model using the R default, the effects parameterization (I do hope that this terminology is not too confusing here …):

```
model.matrix(lm(mass ~ pop + svl))    # Additive model

> model.matrix(lm(mass ~ pop + svl))
  (Intercept)   pop2  pop3   svl
1           1      0     0    40
2           1      0     0    45
```

```
3              1     1     0    39
4              1     1     0    50
5              1     0     1    52
6              1     0     1    57
[ ... ]
```

The first parameter, the intercept, signifies the expected mass in population 1 at the point where the value of the covariate is equal to zero, i.e., for a snake of length zero. That is, therefore, the intercept of the regression model. The parameters associated with the design matrix columns named pop2 and pop3 quantify the difference in the intercept between these populations and population 1. The parameter associated with the last column in the design matrix measures the common slope of mass on svl for snakes in all three populations.

And here is the design matrix of the interaction-effects ANCOVA model using the R default, the effects parameterization:

```
model.matrix(lm(mass ~ pop * svl))       # Interactive model

> model.matrix(lm(mass ~ pop * svl))
   (Intercept)    pop2    pop3    svl    pop2:svl    pop3:svl
1            1       0       0     40           0           0
2            1       0       0     45           0           0
3            1       1       0     39          39           0
4            1       1       0     50          50           0
5            1       0       1     52           0          52
6            1       0       1     57           0          57
[ ... ]
```

The parameters associated with the first three columns in this design matrix signify the population 1 intercept and the effects of populations 2 and 3, respectively, i.e., the difference in intercepts. The parameter associated with the fourth column, svl, is the slope of the mass–length regression in the first population, and the parameters associated with the last two columns are the differences between the slopes in populations 2 and 3 relative to the slope in population 1.

Now let's see the main-effects model using the means parameterization:

```
model.matrix(lm(mass ~ pop + svl−1))       # Additive model

> model.matrix(lm(mass ~ pop + svl−1))
   pop1   pop2   pop3    svl
1     1      0      0     40
2     1      0      0     45
3     0      1      0     39
4     0      1      0     50
```

```
5   0    0    1    52
6   0    0    1    57
[ ... ]
```

The parameters associated with the first three columns in the design matrix now directly represent the intercepts for each population, while that associated with the fourth column denotes the common slope of the mass-svl relationship.

And finally, let's formulate the interaction-effects model using the means parameterization.

```
model.matrix(lm(mass ~ (pop * svl – 1 – svl)))        # Interactive model

> model.matrix(lm(mass ~ pop * svl–1 – svl))
  pop1 pop2 pop3  pop1:svl   pop2:svl  pop3:svl
1   1    0    0      40          0         0
2   1    0    0      45          0         0
3   0    1    0       0         39         0
4   0    1    0       0         50         0
5   0    0    1       0          0        52
6   0    0    1       0          0        57
[ ... ]
```

This parameterization of the ANCOVA model with interaction between population and svl contains parameters that have the direct interpretation as intercepts and slopes of the three mass-svl relationships (one in each population). We will see this below, where we will also see that in R we can directly fit in an lm() function an object that is a design matrix.

Here is the matrix–vector description of the main-effects ANCOVA model with effects parameterization applied to our snake data set:

$$
\begin{pmatrix} 6 \\ 8 \\ 5 \\ 7 \\ 9 \\ 11 \end{pmatrix} = \begin{pmatrix} 1 & 0 & 0 & 40 \\ 1 & 0 & 0 & 45 \\ 1 & 1 & 0 & 39 \\ 1 & 1 & 0 & 50 \\ 1 & 0 & 1 & 52 \\ 1 & 0 & 1 & 57 \end{pmatrix} * \begin{pmatrix} \alpha \\ \beta_1 \\ \beta_2 \\ \delta \end{pmatrix} + \begin{pmatrix} \varepsilon_1 \\ \varepsilon_2 \\ \varepsilon_3 \\ \varepsilon_4 \\ \varepsilon_5 \\ \varepsilon_6 \end{pmatrix}
$$

And here is the means parameterization of the main-effects ANCOVA:

$$
\begin{pmatrix} 6 \\ 8 \\ 5 \\ 7 \\ 9 \\ 11 \end{pmatrix} = \begin{pmatrix} 1 & 0 & 0 & 40 \\ 1 & 0 & 0 & 45 \\ 0 & 1 & 0 & 39 \\ 0 & 1 & 0 & 50 \\ 0 & 0 & 1 & 52 \\ 0 & 0 & 1 & 57 \end{pmatrix} * \begin{pmatrix} \alpha_1 \\ \alpha_2 \\ \alpha_3 \\ \delta \end{pmatrix} + \begin{pmatrix} \varepsilon_1 \\ \varepsilon_2 \\ \varepsilon_3 \\ \varepsilon_4 \\ \varepsilon_5 \\ \varepsilon_6 \end{pmatrix}
$$

Here is the effects parameterization of the interactive model:

$$
\begin{pmatrix} 6 \\ 8 \\ 5 \\ 7 \\ 9 \\ 11 \end{pmatrix} =
\begin{pmatrix}
1 & 0 & 0 & 40 & 0 & 0 \\
1 & 0 & 0 & 45 & 0 & 0 \\
1 & 1 & 0 & 39 & 39 & 0 \\
1 & 1 & 0 & 50 & 50 & 0 \\
1 & 0 & 1 & 52 & 0 & 52 \\
1 & 0 & 1 & 57 & 0 & 57
\end{pmatrix}
*
\begin{pmatrix} \alpha \\ \beta_1 \\ \beta_2 \\ \delta \\ \gamma_1 \\ \gamma_2 \end{pmatrix}
+
\begin{pmatrix} \varepsilon_1 \\ \varepsilon_2 \\ \varepsilon_3 \\ \varepsilon_4 \\ \varepsilon_5 \\ \varepsilon_6 \end{pmatrix}
$$

And finally the means parameterization of the same model:

$$
\begin{pmatrix} 6 \\ 8 \\ 5 \\ 7 \\ 9 \\ 11 \end{pmatrix} =
\begin{pmatrix}
1 & 0 & 0 & 40 & 0 & 0 \\
1 & 0 & 0 & 45 & 0 & 0 \\
0 & 1 & 0 & 0 & 39 & 0 \\
0 & 1 & 0 & 0 & 50 & 0 \\
0 & 0 & 1 & 0 & 0 & 52 \\
0 & 0 & 1 & 0 & 0 & 57
\end{pmatrix}
*
\begin{pmatrix} \alpha_1 \\ \alpha_2 \\ \alpha_3 \\ \delta_1 \\ \delta_2 \\ \delta_3 \end{pmatrix}
+
\begin{pmatrix} \varepsilon_1 \\ \varepsilon_2 \\ \varepsilon_3 \\ \varepsilon_4 \\ \varepsilon_5 \\ \varepsilon_6 \end{pmatrix}
$$

Now let's use R to fit the main-effects and the interaction-effects models using both the effects parameterization and the means parameterization. First, R's default effects parameterization of the main-effects model:

```
lm(mass ~ pop + sv1)
> lm(mass ~ pop + sv1)

Call:
lm(formula = mass ~ pop + sv1)

Coefficients:
(Intercept)         pop2         pop3          sv1
   -3.43860     -1.49123      0.05263      0.24561
```

So, the intercept is the expected mass of a snake in population 1 that has sv1 equal to zero. Pop2 and pop3 are the differences in the intercept between populations 2 and 3 compared with that in population 1, and the parameter named sv1 measures the slope of the mass-sv1 relationship common to snakes in all populations.

It is instructive to plot the estimates of the relationships for each population under this model (Fig. 6.5).

```
fm <- lm(mass ~ pop + sv1)            # Refit model
plot(sv1, mass, col = c(rep("red", 2), rep("blue", 2), rep("green", 2)))
abline(fm$coef[1], fm$coef[4], col = "red")
abline(fm$coef[1]+ fm$coef[2], fm$coef[4], col = "blue")
abline(fm$coef[1]+ fm$coef[3], fm$coef[4], col = "green")
```

This model assumes that the mass-sv1 relationship differs among populations only in the average level. What we see then is that population 1 (red) hardly differs from population 3 (green), but that snakes in

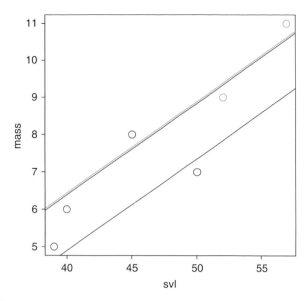

FIGURE 6.5   The main-effects ANCOVA model.

population 2 (blue) weigh less at a given length than do snakes in population 1 or 3.

Next, the interaction-effects models using the R default effects parameterization:

```
lm(mass ~ pop * svl)
> lm(mass ~ pop * svl)

Call:
lm(formula = mass ~ pop * svl)

Coefficients:
(Intercept)       pop2        pop3        svl    pop2:svl    pop3:svl
 -1.000e+01  7.909e+00  -1.800e+00  4.000e-01  -2.182e-01  -6.232e-17
```

The first and the fourth parameters describe intercept and slope of the relationship between mass and svl in the first population, while the remainder refer to intercept and slope differences between the other two populations and those of the baseline population. Note that the last parameter estimate should in fact be zero but is not due to rounding error.

We plot the estimated relationships also under the interaction model (Fig. 6.6):

```
fm <- lm(mass ~ pop * svl)                 # Refit model
plot(svl, mass, col = c(rep("red", 2), rep("blue", 2), rep("green", 2)))
abline(fm$coef[1], fm$coef[4], col = "red")
```

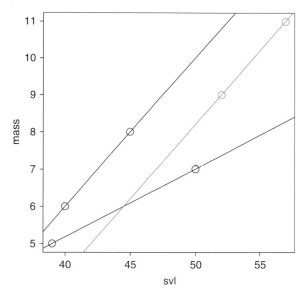

FIGURE 6.6   The interaction-effects ANCOVA model.

```
abline(fm$coef[1]+ fm$coef[2], fm$coef[4] + fm$coef[5], col = "blue")
abline(fm$coef[1]+ fm$coef[3], fm$coef[4] + fm$coef[6], col = "green")
```

As an aside, this is not a useful statistical model, since it has an $R^2 = 1$, but the example does serve to illustrate the two kinds of assumptions about homogeneity or not of slopes that one may examine in an ANCOVA model.

Next, we try the means parameterizations of both the main-effects model and the interaction-effects model. First the main-effects model:

```
lm(mass ~ pop + svl−1)
> lm(mass ~ pop + svl−1)

Call:
lm(formula = mass ~ pop + svl − 1)

Coefficients:
    pop1      pop2      pop3      svl
 −3.4386   −4.9298   −3.3860   0.2456
```

This gives us the estimates of each individual slope plus the slope estimate common to all three populations.

What about the interaction-effects model?

```
lm(mass ~ pop * svl−1 − svl)
> lm(mass ~ pop * svl−1 − svl)
```

```
Call:
lm(formula = mass ~ pop * svl - 1 - svl)

Coefficients:
     pop1        pop2        pop3  pop1:svl  pop2:svl  pop3:svl
 -10.0000     -2.0909    -11.8000    0.4000    0.1818    0.4000
```

These estimates have direct interpretations as the intercept and the slope of the three regressions of mass on svl.

6.4 SUMMARY

We have briefly reviewed the two key components of linear statistical models: statistical distributions and the linear predictor, which is represented by the product of the design matrix and the parameter vector. Understanding both is essential for applied statistics. But while sometimes one may get away in R or other useful stats packages with not exactly knowing what parameterization of a model is fit by the software and what the parameters effectively mean, this is not the case when using Win-BUGS. In WinBUGS, we have to specify all columns of the design matrix and thus must know exactly what parameterization of a model we want to fit and how this is done. The linear models in this chapter were presented in a progression from simple to complex. Chapters 7–11 follow that structure and show how to fit these same models in WinBUGS and also in R for normal responses, i.e., for normal linear models, to which ANOVA, regression, and related methods all belong.

EXERCISE

1. *Fitting a design matrix*: The interaction-effects ANCOVA wasn't a useful statistical model for the toy snake data set, since six fitted parameters perfectly explain six observations and we can't estimate anymore the variability in the system. Use lm( ) to fit a custom-built design matrix, i.e., the design matrix of an ANCOVA with partial interaction effects, where the slopes of the mass–length relationship are the same in population 1 and population 3. Build this design matrix in R, call it X, and fit the model by directly specifying X as the explanatory variable in function lm( ).

t-Test: Equal and Unequal Variances

OUTLINE

One of the most commonly used linear models is that underlying the simple t-test. Actually, almost as with the computation of the arithmetic mean, many people wouldn't even think of a t-test as being a form of linear model. The t-test comes in two flavors; one for the case with equal variances and another for unequal variances. We will look at both in this chapter.

7.1 t-TEST WITH EQUAL VARIANCES

7.1.1 Introduction

The model underlying the t-test with equal variances states that:

$$y_i = \alpha + \beta * x_i + \varepsilon_i$$
$$\varepsilon_i \sim Normal\,(0, \sigma^2)$$

Here, a response y_i is a measurement on a continuous scale taken on individuals i, which belong to either of two groups, and x_i is an indicator or dummy variable for group 2. (See Chapter 6 for different parameterizations of this model.) This simple t-test model has three parameters, the mean α for group 1, the difference in the means between groups 1 and 2 (β) and the variance σ^2 of the normal distribution from which the residuals ε_i are assumed to have come from.

7.1.2 Data Generation

We first simulate data under this model and for a motivating example return to peregrine falcons. We imagine that we had measured male and female wingspan and are interested in a sex difference in size. For Western Europe, Monneret et al. (2006) gives the range of male and female wingspan as 70–85 cm and 95–115 cm, respectively. Assuming normal distributions for wingspan, this implies means and standard deviations of about 77.5 and 2.5 cm for males and 105 and 3 cm for females.

```
n1 <- 60                              # Number of females
n2 <- 40                              # Number of males
mu1 <- 105                            # Population mean of females
mu2 <- 77.5                           # Population mean of males
sigma <- 2.75                         # Average population SD of both

n <- n1+n2                            # Total sample size
y1 <- rnorm(n1, mu1, sigma)           # Data for females
y2 <- rnorm(n2, mu2, sigma)           # Date for males
y <- c(y1, y2)                        # Aggregate both data sets
x <- rep(c(0,1), c(n1, n2))           # Indicator for male
boxplot (y ~ x, col = "grey", xlab = "Male", ylab = "Wingspan (cm)", las = 1)
```

The manner in which we just generated this data set (Fig. 7.1) corresponds in a way to a means parameterization of the linear model of the t-test. Here is a different way to generate a set of the same kind of data. Perhaps it lets one see more clearly the principle of an effects parameterization of the linear model:

```
n <- n1+n2                            # Total sample size
alpha <- mu1                          # Mean for females serves as the intercept
```

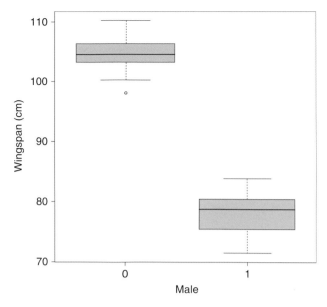

FIGURE 7.1 A boxplot of the generated data set on wingspan of female and male peregrines when the residual variance is constant (0 - females, 1 - males).

```
beta <- mu2-mu1                 # Beta is the difference male-female
E.y <- alpha + beta*x           # Expectation
y.obs <- rnorm(n = n, mean = E.y, sd = sigma)    # Add random variation
boxplot(y.obs ~ x, col = "grey", xlab = "Male", ylab = "Wingspan (cm)", las = 1)
```

An important aside (again): To get a feel for the effect of chance, or technically, for sampling variance (= sampling error), you can repeatedly execute one of the previous sets of commands and observe how different repeated realizations of the same random process are.

7.1.3 Analysis Using R

There is an R function called t.test(), but we will use the linear model function lm() instead to fit the t-test with equal variances for both groups.

```
fit1 <- lm(y ~ x)          # Analysis of first data set
fit2 <- lm(y.obs ~ x)      # Analysis of second data set
summary(fit1)
summary(fit2)

> fit1 <- lm(y ~ x)
> fit2 <- lm(y.obs ~ x)
> summary(fit1)

[...]
```

```
Coefficients:
              Estimate   Std. Error   t value   Pr(>|t|)
(Intercept) 104.6452        0.3592    291.37     <2e-16 ***
x           −26.5737        0.5679    −46.80     <2e-16 ***
[...]
Residual standard error: 2.782 on 98 degrees of freedom
Multiple R-squared: 0.9572,     Adjusted R-squared: 0.9567
F-statistic: 2190 on 1 and 98 DF,    p-value: < 2.2e-16

> summary(fit2)

[...]

Coefficients:
              Estimate   Std. Error   t value   Pr(>|t|)
(Intercept) 105.1785        0.3621    290.45     <2e-16 ***
x           −27.6985        0.5726    −48.38     <2e-16 ***
[...]
Residual standard error: 2.805 on 98 degrees of freedom
Multiple R-squared: 0.9598,     Adjusted R-squared: 0.9594
F-statistic:  2340 on 1 and 98 DF,  p-value: < 2.2e-16
```

The difference between the two analyses is because of sampling variance, i.e., the fact that two different samples from the same population were taken. You may use anova() for an analysis of variance (ANOVA) table of this model:

```
anova(fit1)
anova(fit2)
```

Just for fun check the design matrices for the two models (they are the same):

```
model.matrix(fit1)
model.matrix(fit2)
```

7.1.4 Analysis Using WinBUGS

Here's the Bayesian analysis of the first data set (and don't forget to load package R2WinBUGS). We need to specify the model first. As an extra, we also compute the residuals.

```
# Define BUGS model
sink("ttest.txt")
cat("
model {

# Priors
 mu1 ~ dnorm (0,0.001)            # Precision = 1/variance
```

```
delta ~ dnorm (0,0.001)                # Large variance = Small precision
tau <- 1/ (sigma * sigma)
sigma ~ dunif(0, 10)

# Likelihood
 for (i in 1:n) {
    y[i] ~ dnorm(mu[i], tau)
    mu[i] <- mu1 + delta *x[i]
    residual[i] <- y[i] - mu[i]        # Define residuals
 }

# Derived quantities: one of the greatest things about a Bayesian analysis
 mu2 <- mu1 + delta            # Difference in wingspan
 }
 ",fill=TRUE)
sink()
```

Then, we provide the data, a function to generate inits, a list of parameters we want WinBUGS to keep track of, and specify the Markov chain Monte Carlo (MCMC) settings, after which, we use the function bugs() to run the analysis. Note the use of lognormal (instead of normal) random numbers as initial values for the positive-valued standard deviation sigma.

```
# Bundle data
win.data <- list("x", "y", "n")

# Inits function
inits <- function(){list(mu1=rnorm(1), delta=rnorm(1), sigma = rlnorm(1))}

# Parameters to estimate
params <- c("mu1","mu2", "delta", "sigma", "residual")

# MCMC settings
nc <- 3              # Number of chains
ni <- 1000           # Number of draws from posterior for each chain
nb <- 1              # Number of draws to discard as burn-in
nt <- 1              # Thinning rate

# Start Gibbs sampler
out <- bugs(data = win.data, inits = inits, parameters = params, model =
"ttest.txt", n.thin=nt, n.chains=nc, n.burnin=nb, n.iter=ni, debug = TRUE,
working.directory = getwd())

print(out, dig = 3)

> print(out, dig = 3)
Inference for Bugs model at "ttest.txt", fit using WinBUGS,
 3 chains, each with 1000 iterations (first 1 discarded)
 n.sims = 2997 iterations saved
```

```
                mean     sd     2.5%      25%      50%      75%    97.5%  Rhat   n.eff
mu1          104.638  0.367  103.900  104.400  104.600  104.900  105.400  1.001   2100
mu2           78.070  0.448   77.200   77.770   78.080   78.370   78.950  1.001   3000
delta        -26.568  0.577  -27.680  -26.960  -26.560  -26.190  -25.410  1.001   3000
sigma          2.820  0.205    2.471    2.672    2.808    2.950    3.250  1.001   2400
residual[1]   -2.671  0.366   -3.411   -2.911   -2.669   -2.430   -1.937  1.001   2200
[ ...]
residual[100]  4.579  0.449    3.701    4.279    4.573    4.875    5.443  1.001   3000
deviance     489.489  2.579  486.600  487.700  488.800  490.600  496.300  1.001   3000

[ ...]

DIC info (using the rule, pD = Dbar-Dhat)
pD = 3.0 and DIC = 492.5
DIC is an estimate of expected predictive error (lower deviance is better).
>
```

Comparing the inference from WinBUGS with that using frequentist statistics, we see that the means estimates are almost identical, but that the residual standard deviation estimate is slightly larger in WinBUGS. This last point is general. In later chapters, we will often see that estimates of variances are greater in a Bayesian than in a frequentist analysis. Presumably, the difference will be greatest with smaller sample sizes. This is an indication of the approximate and asymptotic nature of frequentist inference that may differ from the exact inference under the Bayesian paradigm.

One of the nicest things about a Bayesian analysis is that parameters that are functions of primary parameters and their uncertainty (e.g., standard errors or credible intervals) can very easily be obtained using the MCMC posterior samples. Thus, in the above model code, the primary parameters are the female mean and the male–female difference, but we just added a line that computes the mean for males at every iteration, and we directly obtain samples from the posterior distributions of not only the female mean wingspan and the sex difference, but also directly of the mean male wingspan. In a frequentist mode of inference, this would require application of the delta method which is more complicated and also makes more assumptions. In the Bayesian analysis, estimation error is automatically propagated into functions of parameters.

Here are two further comments about model assessment. First, we see that the effective number of parameters estimated is 3.0, which is right because we are estimating one variance and two means. And second, before making an inference about the wingspan in this peregrine population, we should really check whether the model is adequate. Of course, the check of model adequacy is somewhat contrived because we use exclusively simulated and therefore, in a sense, perfect data sets. However, it is important to practice, so we will check the residuals here.

One of the first things to notice about the residuals, which is a little strange at first, is that each one has a distribution. This makes sense, because in a Bayesian analysis, every unknown has a posterior distribution representing our uncertainty about the magnitude of that unknown. Here, we plot the residuals against the order in which individuals were present in the data set and then produce a boxplot for male and female residuals to get a feel whether the distributions of residuals for the two groups are similar.

```
plot(1:100, out$mean$residual)
abline(h = 0)

boxplot(out$mean$residual ~ x, col = "grey", xlab = "Male", ylab = "Wingspan
residuals (cm)", las = 1)
abline(h = 0)
```

No violation of the model assumption of homoscedasticity is apparent from these residual checks.

7.2 t-TEST WITH UNEQUAL VARIANCES

7.2.1 Introduction

The previous analysis assumed that interindividual variation in wingspan is identical for male and female peregrines. This may well not be the case and it may be better to use a model that can accommodate possibly different variances. Our model then becomes as follows:

$$y_i = \alpha + \beta * x_i + \varepsilon_i$$
$$\varepsilon_i \sim Normal(0, \sigma_1^2) \text{ for } x_i = 0 \text{ (females)}$$
$$\varepsilon_i \sim Normal(0, \sigma_2^2) \text{ for } x_i = 1 \text{ (males)}$$

7.2.2 Data Generation

We first simulate data under the heterogeneous groups model (Fig. 7.2).

```
n1 <- 60              # Number of females
n2 <- 40              # Number of males
mu1 <- 105            # Population mean for females
mu2 <- 77.5           # Population mean for males
sigma1 <- 3           # Population SD for females
sigma2 <- 2.5         # Population SD for males

n <- n1+n2            # Total sample size
y1 <- rnorm(n1, mu1, sigma1)    # Data for females
```

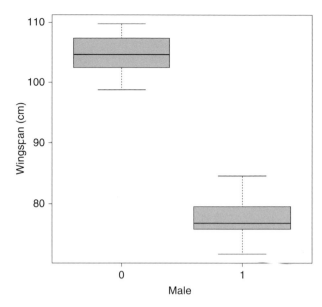

FIGURE 7.2   A boxplot of the generated data set on wingspan of female and male peregrines when the residuals depend on sex (0 - females, 1 - males).

```
y2 <- rnorm(n2, mu2, sigma2)        # Data for males
y <- c(y1, y2)                      # Aggregate both data sets
x <- rep(c(0,1), c(n1, n2))         # Indicator for male
boxplot(y ~ x, col = "grey", xlab = "Male", ylab = "Wingspan (cm)", las = 1)
```

7.2.3 Analysis Using R

A frequentist analysis, using the Welch test to allow for unequal variances, is easy. R defaults to heterogeneous variances when calling the t-test function:

```
t.test(y ~ x)
> t.test(y ~ x)

        Welch Two Sample t-test

data: y by x
t = 46.7144, df = 88.497, p-value < 2.2e-16
alternative hypothesis: true difference in means is not equal to 0
95 percent confidence interval:
 25.93319  28.23750
sample estimates:
mean in group 0   mean in group 1
     104.67641          77.59106
```

7.2.4 Analysis Using WinBUGS

Now here is the Bayesian analysis.

```
# Define BUGS model
sink("h.ttest.txt")
cat("
model {

# Priors
 mu1 ~ dnorm(0,0.001)
 mu2 ~ dnorm(0,0.001)
 tau1 <- 1 / ( sigma1 * sigma1)
 sigma1 ~ dunif(0, 1000)            # Note: Large var. = Small precision
 tau2 <- 1 / ( sigma2 * sigma2)
 sigma2 ~ dunif(0, 1000)

# Likelihood
 for (i in 1:n1) {
    y1[i] ~ dnorm(mu1, tau1)
 }

 for (i in 1:n2) {
    y2[i] ~ dnorm(mu2, tau2)
 }

# Derived quantities
 delta <- mu2 - mu1
 }
 ",fill=TRUE)
sink()

# Bundle data
win.data <- list("y1", "y2", "n1", "n2")

# Inits function
inits <- function(){ list(mu1=rnorm(1), mu2=rnorm(1), sigma1 = rlnorm(1), sigma2 =
rlnorm(1))}

# Parameters to estimate
params <- c("mu1","mu2", "delta", "sigma1", "sigma2")

# MCMC settings
nc <- 3             # Number of chains
ni <- 2000          # Number of draws from posterior for each chain
nb <- 500           # Number of draws to discard as burn-in
nt <- 1             # Thinning rate
```

```
# Unleash Gibbs sampler
out <- bugs(data = win.data, inits = inits, parameters = params, model =
"h.ttest.txt", n.thin = nt, n.chains = nc, n.burnin = nb, n.iter = ni, debug =
TRUE)

print(out, dig = 3)
> print(out, dig = 3)
Inference for Bugs model at "h.ttest.txt", fit using WinBUGS,
 3 chains, each with 2000 iterations (first 500 discarded)
 n.sims = 4500 iterations saved
              mean      sd    2.5%     25%     50%     75%    97.5%  Rhat  n.eff
mu1        104.667   0.400  103.900 104.400 104.700 104.900 105.400 1.001  4500
mu2         77.573   0.444   76.705  77.270  77.570  77.870  78.460 1.001  2700
delta      -27.093   0.592  -28.250 -27.490 -27.090 -26.700 -25.890 1.001  4500
sigma1       3.053   0.292    2.545   2.848   3.029   3.236   3.688 1.001  4500
sigma2       2.825   0.328    2.269   2.592   2.795   3.025   3.540 1.001  4500
deviance   497.784   2.882  494.200 495.600 497.200 499.300 504.952 1.001  4500

[ ... ]

DIC info (using the rule, pD = Dbar-Dhat)
pD = 3.9 and DIC = 501.7
DIC is an estimate of expected predictive error (lower deviance is better).
```

The complication of sex-dependent variances is trivial to deal with in the Bayesian framework. To formally test whether the two variances really differ, one could reparameterize the model such that the variance for one group is expressed as the variance of the other plus some constant to be estimated. Actually, this could also be done "outside" of WinBUGS in R by forming the difference, for each draw in the Markov chain, between sigma1 and sigma2. If the credible interval for that parameter covers zero, then that would be taken as lack of evidence for different variances. This is an important idea; that derived variables with their full posterior uncertainty can also be computed outside of WinBUGS in R if posterior samples of all of their components are available. This is often easier than putting the added code into the WinBUGS model description.

7.3 SUMMARY AND A COMMENT ON THE MODELING OF VARIANCES

We have used WinBUGS to conduct the most widely used statistical test, the t-test. The version of that test with unequal variances is the only place in this book where we explicitly model the variance (except for the modeling of variances by variance components; see Chapters 9, 12, 16, 19, 20, and 21). This chapter shows that not only the mean but also the variance may be modeled. In classical statistics, variance modeling may be rather hard and fairly obscure in its application to an ecologist. In contrast in WinBUGS, the modeling of variances, e.g., as a function of some covariate, could be simply undertaken by use of a log link function; see Lee and Nelder (2006) and Lee et al. (2006) for (frequentist) examples of such models and Exercise 4 (below) for a Bayesian example. Variance modeling, either for the residuals or for random effects, may be required to adequately characterize the stochastic system components when inference is focusing on the mean structure. Alternatively, one may focus on a relation between an explanatory variable and a variance, for instance, to test a hypothesis that some conditions increase the variance in some trait.

EXERCISES

1. *Comparing variances*: See whether you can adapt the WinBUGS code directly to test for equal variances of wingspan. Try a solution using the quantities monitored in the previous analysis.

2. *Assumption violations*: Use simulation to study the effects of heterogeneous variances on the inference by a t-test that assumes homogeneous variances. Assemble data with different SD for males and females, but a common mean, and analyze them in WinBUGS assuming a common dispersion. See what kind of bias is introduced.

3. *Swiss hare data 1*: Use WinBUGS to fit a t-test to the *mean.density* in arable and grassland sites. Repeat that assuming unequal variances. Test for a difference of the variance.

4. *Swiss hare data 2*: Use WinBUGS to fit a t-test to the *mean.density* in arable and grassland sites and introduce a log-linear regression of the variance on *elevation*.

CHAPTER

8

Normal Linear Regression

8.1 INTRODUCTION

We have seen in Chapter 6 that the linear model underlying the simple normal linear regression is the same as that for the t-test:

$$y_i = \alpha + \beta * x_i + \varepsilon_i$$
$$\varepsilon_i \sim Normal\,(0, \sigma^2)$$

The only difference is that the variable x_i doesn't just take on two possible values to indicate membership to one of two groups; rather, it is a measurement that can take on any possible value, within some bounds and up to measurement accuracy. The geometric representation of this model is a straight line, with α being the intercept and β the slope.

As a motivating example for a linear regression analysis, we take a Swiss survey of the wallcreeper (Fig. 8.1), a spectacular little cliff-inhabiting bird that appears to have declined greatly in Switzerland in recent years.

Introduction to WinBUGS for Ecologists
DOI: 10.1016/B978-0-12-378605-0.00008-9

FIGURE 8.1   Wallcreeper (*Tichodroma muraria*), Switzerland, 1989. (*Photo E. Hüttenmoser*)

Assume that we had data on the proportion of sample quadrats in which the species was observed in Switzerland for the years 1990–2005 and that we were willing to assume that the random deviations about a linear time trend were normally distributed. This is for illustration only; usually, we would use logistic regression (Chapters 17–19) or a site-occupancy model (see Chapter 20) to make inference about such data that have to do with the distribution of a species and represent a proportion (i.e., number sites occupied/number sites surveyed).

Importantly, in this chapter, we will also introduce posterior predictive model checking, including the Bayesian *p*-value (Gelman et al., 1996; Gelman and Hill, 2007, Chapter 24). This is a very general concept for checking the goodness-of-fit of a model analysed using simulation techniques like MCMC.

8.2 DATA GENERATION

We generate simple linear regression data (see later Fig. 8.4):

```
n <- 16              # Number of years
a = 40               # Intercept
b = -1.5             # Slope
sigma2 = 25          # Residual variance
```

```
x <- 1:16                        # Values of covariate year
eps <- rnorm(n, mean = 0, sd = sqrt(sigma2))
y <- a + b*x + eps               # Assemble data set
plot((x+1989), y, xlab = "Year", las = 1, ylab = "Prop. occupied (%)", cex = 1.2)
```

8.3 ANALYSIS USING R

Here is a classical analysis using the linear regression facility in R. The function I(.) makes it more straightforward to plot the results:

```
print(summary(lm(y ~ I(x+1989))))
abline(lm(y~ I(x+1989)), col = "blue", lwd = 2)
```

8.4 ANALYSIS USING WinBUGS

Next, we conduct a Bayesian analysis of the same model, which will also include a posterior predictive check plus a Bayesian *p*-value (Gelman et al., 1996) to assess the adequacy of the model for our data set. (Hint: If you don't understand some WinBUGS expressions, such as pow() or step(), open the WinBUGS manual under the Help menu, and in the contents go to Model Specification > Logical nodes.)

8.4.1 Fitting the Model

```
# Write model
sink("linreg.txt")
cat("
model {

# Priors
 alpha ~ dnorm(0,0.001)
 beta ~ dnorm(0,0.001)
 sigma ~ dunif(0, 100)

# Likelihood
 for (i in 1:n) {
    y[i] ~ dnorm(mu[i], tau)
    mu[i] <- alpha + beta*x[i]
 }

# Derived quantities
 tau <- 1/ (sigma * sigma)
 p.decline <- 1-step(beta)        # Probability of decline
```

```
# Assess model fit using a sums-of-squares-type discrepancy
  for (i in 1:n) {
     residual[i] <- y[i]-mu[i]        # Residuals for observed data
     predicted[i] <- mu[i]            # Predicted values
     sq[i] <- pow(residual[i], 2)     # Squared residuals for observed data

# Generate replicate data and compute fit stats for them
     y.new[i] ~ dnorm(mu[i], tau) # one new data set at each MCMC iteration
     sq.new[i] <  pow(y.new[i]-predicted[i], 2) # Squared residuals for new data
  }
  fit <- sum(sq[])                 # Sum of squared residuals for actual data set
  fit.new <- sum(sq.new[])         # Sum of squared residuals for new data set
  test <- step(fit.new - fit)      # Test whether new data set more extreme
  bpvalue <- mean(test)            # Bayesian p-value
  }
", fill=TRUE)
sink()

# Bundle data
win.data <- list("x","y", "n")

# Inits function
inits <- function(){ list(alpha=rnorm(1), beta=rnorm(1), sigma = rlnorm(1))}

# Parameters to estimate
params <- c("alpha","beta", "p.decline", "sigma", "fit", "fit.new", "bpvalue",
"residual", "predicted")

# MCMC settings
nc = 3 ; ni=1200 ; nb=200 ; nt=1

# Start Gibbs sampler
out <- bugs(data = win.data, inits = inits, parameters = params, model =
"linreg.txt", n.thin = nt, n.chains = nc, n.burnin = nb, n.iter = ni, debug = TRUE)

print(out, dig = 3)
```

8.4.2 Goodness-of-Fit Assessment in Bayesian Analyses

In the WinBUGS code, there are two components included to assess the goodness-of-fit of our model. First, there are two lines that compute residuals and predicted values under the model. And second, there is code to compute a Bayesian p-value, i.e., a posterior predictive check (Gelman et al., 1996, 2004; Gelman and Hill, 2007). As an instructive example, we will assess the adequacy of the model using a traditional residual check and then using posterior predictive distributions, including a Bayesian p-value, as an overall measure of fit for a chosen fit criterion.

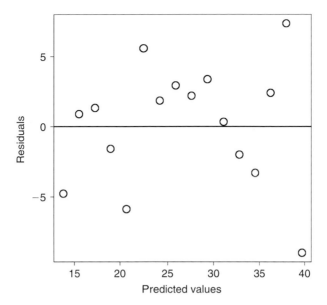

FIGURE 8.2   Residual plot for the linear regression analysis for trend in the Swiss wallcreeper distribution.

Residual Plots

One commonly produced graphical check of the residuals of a linear model is a plot of the residuals against the predicted values. Under the normal linear regression model, residuals are assumed to be a random sample from one single normal distribution. There should be no visible structure in the residuals. In particular, the scatterplot of the residuals should not have the shape of a fan which would indicate that the variance is not constant but is larger, or smaller, for larger responses. We check this first and find no sign of a violation of the homoscedasticity assumption (Fig. 8.2).

```
plot(out$mean$predicted, out$mean$residual, main = "Residuals vs. predicted
values", las = 1, xlab = "Predicted values", ylab = "Residuals")
abline(h = 0)
```

Posterior Predictive Distributions and Bayesian p-Values

The use of posterior predictive distributions is a very general way of assessing the fit of a model when using MCMC model fitting techniques (Gelman et al., 1996; Gelman and Hill, 2007). The idea of a posterior predictive check is to compare the lack of fit of the model for the actual data set with the lack of fit of the model when fitted to replicated, "ideal" data sets. Ideal means that a data set conforms exactly to the assumptions made by the model and is generated under the parameter estimates obtained from the analysis of the actual data set. In contrast to a frequentist analysis, where the solution of a model consists in a single value for each parameter, we estimate a distribution in a Bayesian analysis; hence, any lack-of-fit statistic will also have a distribution.

To obtain such perfect data sets, at each MCMC iteration one replicate data set is assembled under the same model that we fit to the actual data set and using the values of all parameters from the current MCMC iteration. A discrepancy measure chosen to embody a certain kind of lack of fit is computed for both that perfect data set and for the actual data set. Therefore, at the end of an MCMC run for n chains of length m, we have $n*m$ draws from the posterior predictive distribution of the discrepancy measure applied to the actual data set as well as for the discrepancy measure applied to a perfect data set.

What does "discrepancy measure" mean and how is it chosen? The discrepancy measure can be chosen to assess particular features of the model. Often, some global measure of lack of fit will be selected, e.g., a sums of squares-type of discrepancy as we do here, or a Chi-squared-type discrepancy (see Chapter 21 for another example in a more complex hierarchical model). However, entirely different measures may also be chosen; for instance, a discrepancy measure that quantifies the incidence or magnitude of extreme values to assess the adequacy of the model for outliers; see Gelman et al. (1996) for examples.

One of the best ways to assess model adequacy based on posterior predictive distributions is graphically, in a plot of the lack of fit for the ideal data vs. the lack of fit for the actual data (Fig. 8.3). If the model fits the data, then about half of the points should lie above and half of them below a 1:1 line. Alternatively, a numerical summary, called a Bayesian p-value, can be computed that quantifies the proportion of times when the discrepancy measure for the perfect data sets is greater than the discrepancy measure computed for the actual data set. A fitting model has a Bayesian p-value near 0.5, and values close to 0 or close to 1 suggest doubtful fit of the model.

```
lim <- c(0, 3200)
plot(out$sims.list$fit, out$sims.list$fit.new, main = "Graphical posterior
predictive check", las = 1, xlab = "SSQ for actual data set", ylab = "SSQ for ideal
(new) data sets", xlim = lim, ylim = lim)
abline(0, 1)

mean(out$sims.list$fit.new > out$sims.list$fit) # Bayesian p-value
> mean(out$sims.list$fit.new > out$sims.list$fit)
[1] 0.547
```

The graphical posterior predictive check and the numerical Bayesian p-value concur in suggesting that our fitted model is adequate for the

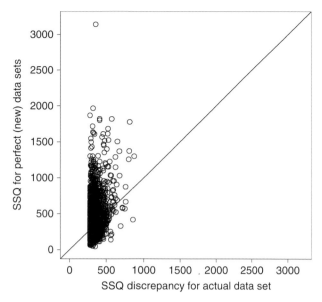

FIGURE 8.3 Graphical posterior predictive check (PPC) of the model adequacy for the wallcreeper analysis plotting predictive vs. realized sums-of-squares discrepancies. The Bayesian *p*-value is equal to the proportion of plot symbols above the 1:1 line. Note that the truncation on the left is due to the hard minimum provided by the least squares estimate.

wallcreeper data, something that will hardly come as a surprise with these simulated data.

Some statisticians don't like posterior predictive checks because they use the data twice: first, to generate the replicate data and second to compare them with these replicates. Model fit assessments based on posterior predictive checks are somewhat too liberal, and posterior predictive checks should not be used for model selection; see Chapter 10 in Ntzoufras (2009) for alternatives.

8.4.3 Forming Predictions

Predictions are expected values of the response variable at some hypothetical values of one or more explanatory variables. Forming predictions is extremely important in applied statistical modeling for two reasons. First, predictions, especially when represented as a graph, are one of the best ways of communicating what can be learned from a model. Second, especially for more complex models, for instance, when there are polynomial terms or interactions and also for Poisson or binomial models (see Chapters 13–21), predictions may be the only way to understand what a model is telling us. Of course, for the simple normal straight-line model in this chapter, we can simply look at magnitude and sign of the slope estimate to understand what the model is telling us about the population trend in Swiss wallcreepers. However, as an exercise, we next plot the estimated trend line from the classical (maximum likelihood [ML]) and the Bayesian (MCMC) fit of the linear regression model:

```
plot((x+1989), y, xlab = "Year", las = 1, ylab = "Prop. occupied (%)", cex = 1.2)
abline(lm(y~ I(x+1989)), col = "blue", lwd = 2)
pred.y <- out$mean$alpha + out$mean$beta * x
points(1990:2005, pred.y, type = "l", col = "red", lwd = 2)
text(1994, 20, labels = "blue - ML; red - MCMC", cex = 1.2)
```

Given the small sample size, we get remarkably similar and indeed virtually identical inferences under the two paradigms (Fig. 8.4).

We can also easily calculate a 95% uncertainty interval by simulation. There are two kinds of such an uncertainty interval, one for the actual data set (called a credible interval) and another for a new data set sampled from the same population (called a prediction interval). Because there is more uncertainty about the line when adding the variability due to the new sample, the latter is wider than the former. Here, we give an example of a credible interval. To produce the interval, we compute the expected response for each of the 3000 elements of our posterior sample of the intercept α and the slope β, and at each point along the x-axis (i.e., for each of

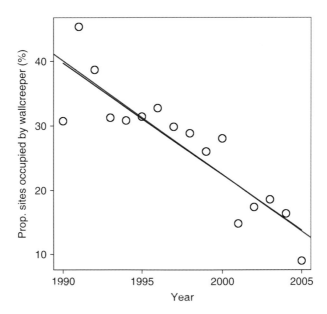

FIGURE 8.4 Observed and predicted change in the distribution of Swiss wallcreepers (blue – maximum likelihood, red – Bayesian).

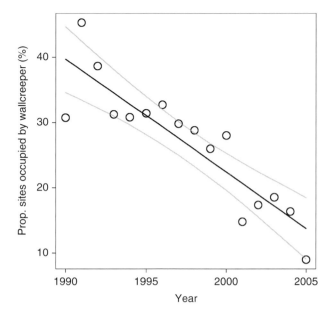

FIGURE 8.5   Predicted wallcreeper trend (black line) with 95% credible interval (grey lines).

the 16 years). Then, we use the 2.5th and 97.5th percentiles of each posterior distribution as our bounds of the credible interval.

We set up an R data structure to hold the predictions, fill them, then determine the appropriate percentile points and produce a plot (Fig. 8.5):

```
predictions <- array(dim = c(length(x), length(out$sims.list$alpha)))
for(i in 1:length(x)){
    predictions[i,] <- out$sims.list$alpha + out$sims.list$beta*i
}
LPB <- apply(predictions, 1, quantile, probs = 0.025) # Lower bound
UPB <- apply(predictions, 1, quantile, probs = 0.975) # Upper bound

plot((x+1989), y, xlab = "Year", las = 1, ylab = "Prop. occupied (%)", cex = 1.2)
points(1990:2005, out$mean$alpha + out$mean$beta * x, type = "l", col = "black",
lwd = 2)
points(1990:2005, LPB, type = "l", col = "grey", lwd = 2)
points(1990:2005, UPB, type = "l", col = "grey", lwd = 2)
```

8.4.4 Interpretation of Confidence vs. Credible Intervals

Consider the frequentist inference about the slope parameter; −1.763, SE 0.241. A quick and dirty frequentist 95% confidence interval is provided by −1.763 ± 2*0.241 = (−2.245, −1.281). This means that if we took, for example, 100 replicate sample observations of 16 annual surveys each in the same Swiss wallcreeper population and 100 times estimated an annual trend with associated 95% CI using linear regression, then on average we would expect 95 intervals would indeed contain the true value of the population trend. We cannot make any direct probability statement about the trend itself; the true value of the trend is either in or out of our single interval, but there is no probability associated with this. In particular, it is wrong to say that the population trend of the wallcreeper lies between −2.245 and −1.281 with a probability of 95%. The probability statement in the 95% CI refers to the reliability of the tool, i.e., computation of the confidence interval, and not to the parameter for which a CI is constructed.

In contrast, the posterior probability in a Bayesian analysis measures our degree of belief about the likely magnitude of a parameter, given the model, the observed data, and our priors. Hence, we can make direct probability statements about a parameter using its posterior distribution. Let's do this here for the slope parameter, which represents the population trend of the wallcreeper in Switzerland (Fig. 8.6).

```
hist(out$sims.list$beta, main = "", col = "grey", xlab = "Trend estimate", xlim =
c(-4, 0))
abline(v = 0, col = "black", lwd = 2)
```

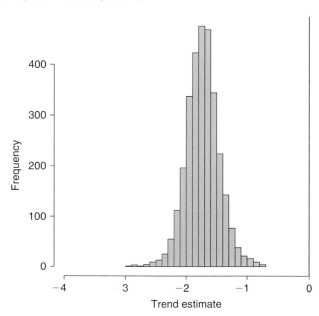

FIGURE 8.6   Posterior distribution of the distributional trend in Swiss wallcreepers. The value of zero (representing no trend) is shown as a black vertical line.

We see clearly that values representing no decline or an increase, i.e., values of the slope of 0 and larger have no mass at all under this posterior distribution. We can thus say that the probability of a stable or increasing wallcreeper population is essentially nil. Such a statement is exactly what most users of statistics, such as politicians, would like to have, rather than the somewhat contorted statement about a population trend as based on the frequentist confidence interval.

8.5 SUMMARY

We have used WinBUGS to fit a linear regression, the algebraic model and the WinBUGS code for which is essentially identical to that for the t-test. We have also introduced posterior predictive distributions along with the Bayesian p-value as a very general and flexible way of assessing goodness-of-fit of a model analyzed using MCMC.

EXERCISES

1. *Toy problem*: Assume you examined five frogs that weighed 10, 20, 23, 32, and 35 g and had lengths of 5, 7, 10, 12, and 15 units. Write out the linear regression model using vectors and matrices and the set of equations implied. Conduct a normal linear regression analysis for this data set using R and WinBUGS.

2. *Prediction*: One way of forming predictions in WinBUGS is by specifying them as additional (derived) variables in the model. Another way is to form them outside of WinBUGS in R, if Markov chains for all the required ingredients are available. The third and perhaps the simplest way is by adding to the data set missing values (NAs) in the response and the desired levels of the explanatory variables in the model. In a Bayesian analysis, missing values are treated exactly like parameters, i.e., WinBUGS will draw samples for each missing value as part of the model fitting. The resulting posterior predictive distribution can then be summarized for inference in the usual way. Try this for the frog data set and predict mass at length 16, 17, 18, 19, and 20 units.

3. *Swiss hare data*: Fit a normal linear regression analysis for *mean.density* on *year* for grassland areas. Hint: You must first select the data for grassland areas and then aggregate over sites to obtain mean annual density in grassland. Does a linear trend adequately capture the variability in the data?

CHAPTER

9

Normal One-Way ANOVA

9.1 INTRODUCTION: FIXED AND RANDOM EFFECTS

Analysis of variance (ANOVA) is the generalization of a t-test to more than two groups. There are different kinds of ANOVA: one-way, with just a single factor, and two- or multiway, with two or more factors, and main- and interaction-effects models (see Chapter 10). Here, we present a one-way ANOVA and introduce the concept of random effects along the way. In random-effects models, a set of effects (e.g., group means) is constrained to come from some distribution, which is most often a normal, although it may be a Bernoulli (see Chapter 20), a Poisson (see Chapter 21) or yet another distribution. In this chapter, we will first generate and analyze a fixed-effects and then a random-effects ANOVA data set. In Chapters 12, 16, and 19–21, we will focus on mixed models, i.e., those containing both fixed and random effects. As a motivating example for this chapter, we assume that we measured snout–vent length

Introduction to WinBUGS for Ecologists
DOI: 10.1016/B978-0-12-378605-0.00009-0
115

FIGURE 9.1   Smooth snake (*Coronella austriaca*), France, 2006. (*Photo C. Berney*)

(SVL) in five populations of Smooth snakes (Fig. 9.1) and were interested in whether populations differ.

The one-way ANOVA can be parameterized in various ways (see Section 6.3.4). We adopt a means parameterization of the linear model for the fixed-effects, one-way ANOVA:

$$y_i = \alpha_{j(i)} + \varepsilon_i$$
$$\varepsilon_i \sim Normal\,(0,\sigma^2)$$

Here, y_i is the observed SVL of Smooth snake i in population j, $\alpha_{j(i)}$ is the expected SVL of a snake in population j, and residual ε_i is the random SVL deviation of snake i from its population mean $\alpha_{j(i)}$. It is assumed to be normally distributed around zero with constant variance σ^2.

Without any further assumption, the population means $\alpha_{j(i)}$ are simply some unknown constants that are estimated in a fixed-effects ANOVA. If, however, we add a distributional assumption about the population means $\alpha_{j(i)}$, we obtain a random-effects ANOVA:

$$y_i = \alpha_{j(i)} + \varepsilon_i$$
$$\varepsilon_i \sim Normal\,(0,\sigma^2)$$
$$\alpha_{j(i)} \sim Normal\,(\mu,\tau^2)$$

The interpretation of $\alpha_{j(i)}$ and ε_i as population mean SVL and residual, respectively, is unchanged. But now, the $\alpha_{j(i)}$ parameters are no longer assumed to be independent; rather, they come from a second normal distribution with mean μ and variance τ^2. The latter are also called hyper-parameters, because they are one level higher than the parameters $\alpha_{j(i)}$ that they govern.

Thus, the models for the fixed- and random-effects ANOVA differ only subtly; so how do we know when to apply which one in practice? Unfortunately, there are fairly differing views on this decision, see Gelman and Hill (2007, p. 245). The traditional view goes about as follows. When you have a particular interest in the studied factor levels and/or when you have included (nearly) all conceivable levels in a study, the associated factor should be viewed as having fixed effects. You estimate the effects of each level but are not interested in the variance among levels, except as part of an ANOVA table to construct an F-test statistic for that factor. Importantly, you cannot generalize to factor levels that you did not study. In contrast, you consider a factor as random when you don't have a particular interest in the levels that actually appear in your study and/or when these levels form a sample from a (much) larger set of possible levels that you *could* have included in your study. Typically, you want to generalize to this larger population and are more interested in the variation among the factor levels in that population, although you may still want to estimate the effects of the levels actually observed in your study. Thus, typical fixed-effects factors would be sex or cereal variety in an agricultural experiment. Typical random-effects factors might be time (e.g., year, month, or day) or location, such as experimental blocks or other spatial units on which repeated measurements are taken.

It has been argued (e.g., Robinson, 1991), that it doesn't make sense to claim that you studied only a sample of effects and then estimate them anyway, and that a more natural distinction between fixed and random effects is simply based on the question of whether they could plausibly have come from some *distribution* of effects. Such random effects are generated by the same homogeneous, stochastic process and statisticians also say they are *exchangeable*, which in common language means that they are similar, but not identical. This similarity is because of the common stochastic process that generated them and thus creates a stochastic relationship among the effects of the levels of a random-effects factor. In contrast, when factor levels are modeled as fixed they are considered unrelated or independent.

Why should one make distributional assumptions about a set of effects in a model, i.e., go from a fixed-effects ANOVA to the corresponding random-effects ANOVA? There are three reasons: extrapolation of inference to a wider population, improved accounting for system uncertainty, and efficiency of estimation. First, viewing the studied effects as a random sample from some population enables one to extrapolate to that population. This generalization can only be achieved by modeling the process that generates the realized values of the random effects (i.e., by assuming a normal distribution for the $\alpha_{j(i)}$ above). Second, declaring factor effects as random acknowledges that when repeating our study, we obtain a different set of effects, so the resulting parameter estimates will differ from those in our current study. Random-effects modeling properly accounts for this added uncertainty in our inference about the analyzed system. Third, when making a random-effects assumption about a factor, these effects are no longer estimated independently; instead, estimates are influenced by each other and therefore are dependent. Specifically, individual estimates are "pulled in" toward the common mean μ, i.e., they are closer to μ than the corresponding fixed-effects estimates. This is why random-effects estimators are said to be "shrinkage estimators". Estimates that are more imprecise and are based on a smaller sample size are shrunk more. When effects are indeed exchangeable, shrinkage results in better estimates (e.g., with smaller prediction error) than the estimates obtained from a fixed-effects analysis. This is why one also says that a random-effects analysis "borrows strength".

Random-effects modeling can also be viewed as a compromise between assuming no effects and fully independent effects of the levels of a factor. When assuming a factor has no effect, you pool its effects, whereas when assuming it has fixed effects, you treat all effects as completely independent instead. When assuming a factor has random effects, you pool effects only partially, and the degree of pooling is based on the amount of information that you have about the effect of each level. According to this view, you might always want to assume all factors as random and let the data determine the degree of pooling; see Gelman (2005) and Gelman and Hill (2007).

Recently, statisticians seem to prefer the view of random-effects factors as those, whose effects result from a common stochastic process, with the resulting benefits of the ability to extrapolate, more honest accounting for uncertainty and shrinkage estimation. For instance, Sauer and Link (2002) assessed population trends in a large numbers of bird species and showed how imprecise estimates for species with little information borrowed strength from the "ensemble" (the group of species) and got pulled toward the group mean, and this yielded better predictions. Similarly, Welham et al. (2004) analyzed a huge wheat variety testing experiment and treated variety as random. Again, they found that this gave better predictions of future yield than treating variety as fixed.

Next, we generate one data set under a fixed-effects design and another under a random-effects design. We do this in a very "linear model" fashion, i.e., by first specifying a design matrix and choosing parameter values. For the fixed-effects analysis, we will arbitrarily select these values, whereas for the random-effects analysis, we will draw them from a normal distribution with specified hyperparameters. Then, we multiply the design matrix by the parameter vector to obtain the linear predictor, to which we add residuals to obtain the actual measurements.

9.2 FIXED-EFFECTS ANOVA

9.2.1 Data Generation

We assume five groups with 10 snakes measured in each with SVL averages of 50, 40, 45, 55, and 60. This corresponds to a baseline population mean of 50 and effects of populations 2–5 of −10, −5, 5, and 10. We choose a residual standard deviation of SVL of 3 and assemble everything (Fig. 9.2). (Note that %*% denotes matrix multiplication in the R code below.)

```
ngroups <- 5              # Number of populations
nsample <- 10             # Number of snakes in each
pop.means <- c(50, 40, 45, 55, 60)   # Population mean SVL
sigma <- 3                # Residual sd

n <- ngroups * nsample    # Total number of data points
eps <- rnorm(n, 0, sigma) # Residuals
x <- rep(1:5, rep(nsample, ngroups)) # Indicator for population
means <- rep(pop.means, rep(nsample, ngroups))
X <- as.matrix(model.matrix(~ as.factor(x) -1)) # Create design matrix
```

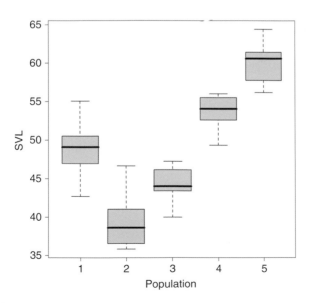

FIGURE 9.2 Snout–vent length (SVL) of Smooth snakes in five populations simulated under a fixed-effects model.

```
X                        # Inspect that
y <- as.numeric(X %*% as.matrix(pop.means) + eps)
      # assemble -- NOTE: as.numeric ESSENTIAL for WinBUGS
boxplot(y~x, col="grey", xlab="Population", ylab="SVL", main="", las = 1)
```

9.2.2 Maximum Likelihood Analysis Using R

```
print(anova(lm(y~as.factor(x))))
cat("\n")
print(summary(lm(y~as.factor(x)))$coeff, dig = 3)
cat("Sigma:     ", summary(lm(y~as.factor(x)))$sigma, "\n")

> print(summary(lm(y~as.factor(x)))$coeff, dig = 3)
              Estimate Std.  Error  t value    Pr(>|t|)
(Intercept)      48.84        0.864   56.51    1.96e-43
as.factor(x)2    -9.48        1.222   -7.75    7.89e-10
as.factor(x)3    -4.54        1.222   -3.71    5.67e-04
as.factor(x)4     4.76        1.222    3.90    3.20e-04
as.factor(x)5    11.25        1.222    9.20    6.56e-12
> cat("Sigma:          ", summary(lm(y~as.factor(x)))$sigma, "\n")
Sigma:                  2.733362
```

Remembering that R fits an effects parameterization, we recognize fairly well the input values of the analysis.

9.2.3 Bayesian Analysis Using WinBUGS

We fit a means parameterization of the model and obtain effects estimates (i.e., population differences) as derived quantities. Note WinBUGS' elegant double-indexing (alpha[x[i]]) to specify the expected SVL of snake i according to the i-th value of the population index x. We also add two lines to show how custom hypotheses can easily be tested as derived quantities. Test 1 examines whether snakes in populations 2 and 3 have the same size as those in populations 4 and 5. Test 2 checks whether the size difference between snakes in populations 5 and 1 is twice that between populations 4 and 1.

```
# Write model
sink("anova.txt")
cat("
model {

# Priors
  for (i in 1:5){        # Implicitly define alpha as a vector
     alpha[i] ~ dnorm(0, 0.001)
  }
  sigma ~ dunif(0, 100)
```

```
# Likelihood
  for (i in 1:50) {
      y[i] ~ dnorm(mean[i], tau)
      mean[i] <- alpha[x[i]]
  }

# Derived quantities
  tau <- 1 / ( sigma * sigma)
  effe2 <- alpha[2] - alpha[1]
  effe3 <- alpha[3] - alpha[1]
  effe4 <- alpha[4] - alpha[1]
  effe5 <- alpha[5] - alpha[1]

# Custom hypothesis test / Define your own contrasts
  test1 <- (effe2+effe3) - (effe4+effe5) # Equals zero when 2+3 = 4+5
  test2 <- effe5 - 2 * effe4 # Equals zero when effe5 = 2*effe4
}
",fill=TRUE)
sink()

# Bundle data
win.data <- list("y", "x")

# Inits function
inits <- function(){ list(alpha = rnorm(5, mean = mean(y)), sigma = rlnorm(1) )}

# Parameters to estimate
params <- c("alpha", "sigma", "effe2", "effe3", "effe4", "effe5", "test1", "test2")

# MCMC settings
ni <- 1200
nb <- 200
nt <- 2
nc <- 3

# Start Gibbs sampling
out <- bugs(win.data, inits, params, "anova.txt", n.thin=nt, n.chains=nc,
n.burnin=nb, n.iter=ni, debug = TRUE)

# Inspect estimates
print(out, dig = 3)
> print(out, dig = 3)
Inference for Bugs model at "anova.txt", fit using WinBUGS,
  3 chains, each with 1200 iterations (first 200 discarded), n.thin = 2
  n.sims = 1500 iterations saved
```

	mean	sd	2.5%	25%	50%	75%	97.5%	Rhat	n.eff
alpha[1]	48.837	0.877	47.049	48.240	48.840	49.420	50.636	1.011	180
alpha[2]	39.354	0.888	37.595	38.770	39.350	39.990	41.080	1.000	1500

```
alpha[3]    44.315  0.867    42.710   43.720    44.290    44.910    46.016   1.002   1500
alpha[4]    53.580  0.886    51.759   53.020    53.565    54.150    55.356   1.001   1500
alpha[5]    60.019  0.882    58.285   59.430    60.030    60.590    61.750   1.001   1500
sigma        2.811  0.303     2.296    2.594     2.777     3.015     3.451   1.002   1200
effe2       -9.482  1.273   -12.040  -10.320    -9.456    -8.665    -7.046   1.003    590
effe3       -4.521  1.275    -7.079   -5.356    -4.548    -3.677    -2.037   1.006    360
effe4        4.744  1.241     2.325    3.944     4.768     5.514     7.205   1.004    470
effe5       11.182  1.239     8.767   10.390    11.180    12.002    13.615   1.002    810
test1      -29.929  1.764   -33.425  -31.100   -29.900   -28.800   -26.415   1.000   1500
test2        1.694  2.158    -2.670    0.319     1.644     3.139     6.016   1.002    940
deviance   243.557  3.671   238.500  240.800   242.800   245.600   252.452   1.005    580
[ ... ]
DIC info (using the rule, pD = Dbar-Dhat)
pD = 5.8 and DIC = 249.3
DIC is an estimate of expected predictive error (lower deviance is better).
```

As an aside, the effective number of parameters, pD, is estimated quite correctly as we have five group means and one variance parameter. Comparison with the maximum likelihood solutions (see Section 9.2.2) shows how with vague priors, a Bayesian analysis yields numerically virtually identical inferences as a frequentist analysis. However, one of the most compelling things about a Bayesian analysis conducted using Markov chain Monte Carlo methods is the ease with which derived quantities can be estimated and custom tests conducted. In the above WinBUGS model code, we see how easily such custom contrasts (comparisons) can be estimated with full error propagation from all the involved random quantities. Of course, for a simple model such as a one-way ANOVA, this can also be done fairly easily in standard stats packages. However, in WinBUGS, this is *equally simple for any kind of parameter*, e.g., for variances (see Chapter 7), *and in any model type*, e.g., mixed models, generalized linear models (GLMs), or generalized linear mixed models (GLMMs).

9.3 RANDOM-EFFECTS ANOVA

9.3.1 Data Generation

For our second data set, we assume that population SVL means come from a normal distribution with selected hyperparameters. The code is only slightly different from the previous data-generating code. First, we choose the two sample sizes; the number of populations and that of snakes examined in each. As always, we generate balanced data for convenience only. The methods to "decompose" a data set in R and WinBUGS work just as well for unbalanced data.

```
npop <- 10                # Number of populations: now choose 10 rather than 5
nsample <- 12             # Number of snakes in each
n <- npop * nsample       # Total number of data points
```

We choose the hyperparameters of the normal distribution from which the random population means are thought to come from and use `rnorm()` to draw one realization from that distribution for each population. Then, we select the residual standard deviation and draw residuals.

```
pop.grand.mean <- 50      # Grand mean SVL
pop.sd <- 5               # sd of population effects about mean
pop.means <- rnorm(n = npop, mean = pop.grand.mean, sd = pop.sd)
sigma <- 3                # Residual sd
eps <- rnorm(n, 0, sigma)  # Draw residuals
```

We build the design matrix, expand the population effects to the chosen (larger) sample size, use matrix multiplication to assemble our data set and have a look at what we've created (Fig. 9.3):

```
x <- rep(1:npop, rep(nsample, npop))
X <- as.matrix(model.matrix(~ as.factor(x) -1))
y <- as.numeric(X %*% as.matrix(pop.means) + eps) # as.numeric is ESSENTIAL
boxplot(y ~ x, col = "grey", xlab = "Population", ylab = "SVL", main = "", las = 1)
                          # Plot of generated data
abline(h = pop.grand.mean)
```

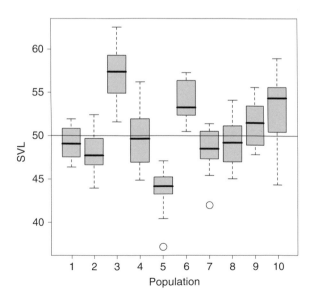

FIGURE 9.3 Simulated snout–vent length (SVL) of Smooth snakes in 10 populations. We can't tell from this graph that the population effects are random now rather than fixed as in Fig. 9.2. The horizontal line shows the grand mean.

9.3.2 Restricted Maximum Likelihood Analysis Using R

The functions contained in the R package lme4 allow one to fit, using approximate methods (Gelman and Hill, 2007), a wide range of linear, nonlinear, and generalized linear mixed models. We use this package for fitting a random-effects ANOVA by restricted maximum likelihood (REML). lme4 requires package Matrix, which we need to download and install in case we haven't done so already.

```
library('lme4')           # Load lme4

pop <- as.factor(x)       # Define x as a factor and call it pop

lme.fit <- lmer(y ~ 1 + 1 | pop, REML = TRUE)
lme.fit                   # Inspect results
ranef(lme.fit)            # Print random effects

> lme.fit# Inspect results
Linear mixed model fit by REML
Formula: y ~ 1 + 1 | pop
   AIC    BIC    logLik   deviance   REMLdev
  630.1  638.5  -312.1      626.3     624.1
Random effects:
   Groups    Name         Variance   Std.Dev.
   pop       (Intercept)  13.1244    3.6228
   Residual                8.5172    2.9184
Number of obs: 120, groups: pop, 10

Fixed effects:
            Estimate   Std. Error   t value
(Intercept) 50.318     1.176        42.78
> ranef(lme.fit)#   Print random effects
$pop
       (Intercept)
1      -1.0784560
2      -2.2585587
3       6.5857470
4      -0.5906953
5      -6.2970554
6       3.3997612
7      -1.9525175
8      -1.0407413
9       0.9989122
10      2.2336038
```

9.3.3 Bayesian Analysis Using WinBUGS

Now, WinBUGS. Note that adopting a common prior distribution for the 10 population means represents the random-effects assumption in this model.

```
sink("re.anova.txt")
cat("
model {

# Priors and some derived things
for (i in 1:npop){
    pop.mean[i] ~ dnorm(mu, tau.group) # Prior for population means
    effe[i] <- pop.mean[i] - mu # Population effects as derived quant's
    }
  mu ~ dnorm(0,0.001)          # Hyperprior for grand mean svl
  sigma.group ~ dunif(0, 10)   # Hyperprior for sd of population effects
  sigma.res ~ dunif(0, 10)     # Prior for residual sd

# Likelihood
  for (i in 1:n) {
    y[i] ~ dnorm(mean[i], tau.res)
    mean[i] <- pop.mean[x[i]]
    }

# Derived quantities
  tau.group <- 1 / (sigma.group * sigma.group)
  tau.res <- 1 / (sigma.res * sigma.res)
}
",fill=TRUE)
sink()

# Bundle data
win.data <- list(y=y, x=x, npop = npop, n = n)

# Inits function
inits <- function(){ list(mu = runif(1, 0, 100), sigma.group = rlnorm(1), sigma.res
= rlnorm(1) )}

# Params to estimate
parameters <- c("mu", "pop.mean", "effe", "sigma.group", "sigma.res")

# MCMC settings
ni <- 1200
nb <- 200
nt <- 2
nc <- 3
```

```
# Start WinBUGS
out <- bugs(win.data, inits, parameters, "re.anova.txt", n.thin=nt, n.chains=nc,
n.burnin=nb, n.iter=ni, debug = TRUE)

# Inspect estimates
print(out, dig = 3)
Inference for Bugs model at "re.anova.txt", fit using WinBUGS,
  3 chains, each with 1200 iterations (first 200 discarded), n.thin = 2
  n.sims = 1500 iterations saved
```

	mean	sd	2.5%	25%	50%	75%	97.5%	Rhat	n.eff
mu	50.199	1.438	47.309	49.337	50.195	51.070	53.076	1.002	1500
pop.mean[1]	49.247	0.834	47.589	48.670	49.250	49.810	50.875	1.001	1500
pop.mean[2]	48.054	0.852	46.405	47.490	48.020	48.640	49.655	1.001	1500
pop.mean[3]	56.933	0.871	55.275	56.340	56.950	57.520	58.590	1.001	1500
pop.mean[4]	49.724	0.839	48.100	49.180	49.710	50.290	51.290	1.001	1500
pop.mean[5]	43.962	0.847	42.290	43.380	43.980	44.540	45.510	1.001	1500
pop.mean[6]	53.705	0.849	52.070	53.120	53.685	54.300	55.370	1.000	1500
pop.mean[7]	48.349	0.799	46.715	47.800	48.360	48.920	49.875	1.001	1500
pop.mean[8]	49.289	0.829	47.650	48.720	49.310	49.842	50.846	1.003	1000
pop.mean[9]	51.267	0.844	49.650	50.700	51.240	51.860	52.920	1.001	1500
pop.mean[10]	52.567	0.806	51.020	52.000	52.580	53.130	54.115	1.004	540
effe[1]	-0.952	1.589	-4.156	-1.990	-0.926	0.095	2.189	1.001	1500
effe[2]	-2.145	1.631	-5.513	-3.197	-2.126	-1.113	1.171	1.002	1200
effe[3]	6.734	1.639	3.558	5.680	6.781	7.736	10.055	1.001	1500
effe[4]	-0.475	1.617	-3.716	-1.461	-0.493	0.532	2.744	1.001	1500
effe[5]	-6.237	1.628	-9.426	-7.231	-6.255	-5.202	-3.169	1.001	1500
effe[6]	3.506	1.616	0.458	2.490	3.435	4.542	6.749	1.001	1500
effe[7]	-1.850	1.576	-5.136	-2.832	-1.858	-0.832	1.124	1.000	1500
effe[8]	-0.910	1.611	-4.105	-1.940	-0.898	0.074	2.352	1.003	1300
effe[9]	1.067	1.622	-1.975	-0.008	1.096	2.121	4.104	1.001	1500
effe[10]	2.368	1.582	-0.729	1.342	2.383	3.392	5.595	1.001	1500
sigma.group	4.287	1.254	2.546	3.380	4.036	4.941	7.305	1.002	1400
sigma.res	2.946	0.199	2.584	2.804	2.931	3.077	3.376	1.000	1500
deviance	598.756	5.010	591.347	595.200	597.900	601.600	610.210	1.005	590

```
[ ... ]
DIC info (using the rule, pD = Dbar-Dhat)
pD = 10.6 and DIC = 609.3
DIC is an estimate of expected predictive error (lower deviance is better).
```

On comparing the inference using REML in lme4 (see Section 9.3.2) with our Bayesian analysis, we find similar, but not identical, results of the two analyses. For instance, the among-population SVL variation (sigma. group), is estimated better in the Bayesian analysis (remember that truth is 5). The classical analysis, using lmer(), appears to be more biased and does not give a standard error of the estimated random-effects standard deviation. This illustrates that `lmer()` may yield more biased estimates for small samples, i.e., when there are few levels only (Gelman and Hill, 2007). However, it has to be said that estimation of a between-group variance is always difficult when there are few levels and/or that variance is small (Lambert et al., 2005).

9.4 SUMMARY

We have introduced fixed (independent) and random (dependent) factor effects and fitted the corresponding one-way ANOVA models using Win-BUGS and R, using lme4. We found that the WinBUGS solution is presumably less biased for the random-effects variance than the solution by lme4 when sample sizes are small.

EXERCISES

1. *Convert the ANOVA to an ANCOVA* (analysis of covariance): Within the fixed-effects ANOVA, add the effect on SVL of a continuous measure of habitat quality that varies by individual snake (perhaps individuals in better habitat are larger and more competitive). You may either recreate a new data set that contains such an effect or simply create a habitat covariate (e.g., by drawing random numbers) and add it as a covariate into the previous analysis.
2. *Watch shrinkage happen*: Population mean estimates under the random-effects ANOVA are shrunk toward the grand mean when compared with those under a fixed-effects model. First, watch this shrinkage by fitting a fixed-effects ANOVA to the random-effects data simulated in Section 9.3.1. Second, discard the data from 8 out of 10 snakes in one population and see what happens to the estimate of that population mean.
3. *Swiss hare data*: Compare mean observed population density among all surveyed sites when treating years as replicates. Do this once assuming that these populations corresponded to fixed effects; then repeat the analysis assuming they are random effects.

Normal Two-Way ANOVA

10.1 INTRODUCTION: MAIN AND INTERACTION EFFECTS

We now extend the one-way analysis of variance (ANOVA) model by adding another factor and arrive at a two-way ANOVA. We only consider fixed effects here. There are two ways in which the effects of two factors A and B can be combined, and the associated models are called main-effects and interaction-effects model. In the main-effects model, the effects of A and B are additive, i.e., the effect of one level of factor A, say a_1, does not depend on whether it is assessed at one level of B, say b_1, or at another, say b_2. In contrast, with an interaction between A and B, some or all effects depend on some or all of each other and the effect of a_1 may not be identical when assessed at b_1 or at b_2. Interaction is symmetric, so the effect of b_1 will in general also not be the same whether assessed at a_1 or at a_2. However, the interaction model is still linear, since effects are simply

Introduction to WinBUGS for Ecologists
DOI: 10.1016/B978-0-12-378605-0.00010-7

added together, only with an additional set of effects: those for the combination of each level of two or more factors. When factors are considered fixed, then depending on the model, not all effects will be estimable, see later (10.5.1). In contrast, in a random-effects model with interaction, all effects will in general be estimable.

In Chapter 6 we already saw the linear models for the two-way ANOVA with interaction. In short, the effects parameterization for a model with two factors A (with j levels) and B (with k levels) is this:

$$y_i = \alpha + \beta_{j(i)} * A_i + \delta_{k(i)} * B_i + \gamma_{jk(i)} * A_i * B_i + \varepsilon_i,$$

while the means parameterization is this:

$$y_i = \alpha_{jk(i)} * A_i * B_i + \varepsilon_i$$

In both cases, we need to assume a distribution for the residuals to complete the model description:

$$\varepsilon_i \sim Normal\ (0, \sigma^2).$$

We will use the beautiful mourning cloak (Fig. 10.1) as an illustration for this chapter and assume that we measured wing length of butterflies in each of three elevation classes in five different populations and that the effects of these factors interact. Table 10.1 shows the meaning of the coefficients in the linear model for the effects parameterization.

FIGURE 10.1   Mourning cloak (*Nymphalis antiopa*), Switzerland, 2006. (*Photo: T. Marent*)

TABLE 10.1 The 15 Parameters Estimated in the Effects Parameterization of a Two-Way ANOVA with Interaction for the Mourning Cloak Example.

Intercept	Elevation2	Elevation3
pop2	pop2.elevation2	pop2.elevation3
pop3	pop3.elevation2	pop3.elevation3
pop4	pop4.elevation2	pop4.elevation3
pop5	pop5.elevation2	pop5.elevation3

If the population factor has n.pop levels and the elevation factor n.elev levels, we have one intercept, n.pop-1 = 4 effects for the population factor, n.elev-1 = 2 effects for the elevation factor and (n.pop-1)* (n.elev-1) = 8 effects for the interaction between population and elevation. This adds up to the 15 degrees of freedom that it will cost us to fit this model to a data set that contains observations in every cell (combination of levels) in the cross-classification of population and elevation.

10.2 DATA GENERATION

We assume five populations with 12 butterflies were measured in each and that of these 12, four butterflies were studied at each of three elevation classes (low, medium, high). Wing length differs with elevation, perhaps because butterflies hatch at different size at different elevation or because of different size-dependent predation owing to different bird communities at different elevations. Furthermore, the relationship between wing length and elevation class is not homogeneous among the five studied populations so there is a population-elevation interaction. Residual wing length standard deviation will be 3.

```
# Choose sample size
n.pop <- 5
n.elev <- 3
nsample <- 12
n <- n.pop * nsample

# Create factor levels
pop <- gl(n = n.pop, k = nsample, length = n)
elev <- gl(n = n.elev, k = nsample / n.elev, length = n)

# Choose effects
baseline <- 40                    # Intercept
pop.effects <- c(-10, -5, 5, 10) # Population effects
elev.effects <- c(5, 10)      # Elev effects
```

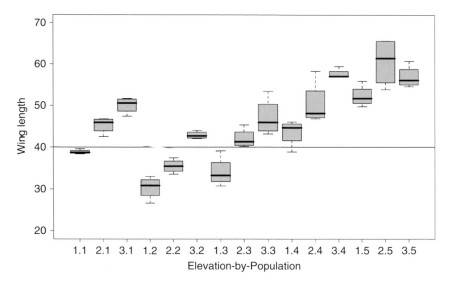

FIGURE 10.2   Mean wing length of mourning cloaks at each of three elevations and in each of five populations. Boxplots are ordered first by elevation and second by population and their identity is recognizable from the tick labels on the x-axis. For instance, the boxplot labelled 3.2 on the x-axis shows the mean wing length in population 2 at elevation 3.

```
interaction.effects <- c(-2, 3, 0, 4, 4, 0, 3, -2)      # Interaction effects
all.effects <- c(baseline, pop.effects, elev.effects, interaction.effects)

sigma <- 3
eps <- rnorm(n, 0, sigma)                # Residuals

X <- as.matrix(model.matrix(~ pop*elev))   # Create design matrix
X                                          # Have a look at that
```

Use matrix multiplication to assemble all components for the final wing length measurements y which we inspect in a grouped boxplot (Fig. 10.2).

```
wing <- as.numeric(as.matrix(X) %*% as.matrix(all.effects) + eps)
     # NOTE: as.numeric is ESSENTIAL for WinBUGS later
boxplot(wing ~ elev*pop, col = "grey", xlab = "Elevation-by-Population", ylab =
"Wing length", main = "Simulated data set", las = 1, ylim = c(20, 70))   # Plot of
generated data
abline(h = 40)
```

We have generated data for which the wing length–elevation relationship varies considerably among the five populations. This can also be nicely seen in a useful conditioning plot, which can be drawn using the function xyplot() in the lattice package. The data can be viewed in two ways; both plots show that the effects of population and elevation are not independent.

```
library("lattice")              # Load the lattice library
xyplot(wing ~ elev | pop, ylab = "Wing length", xlab = "Elevation", main =
"Population-specific relationship between wing and elevation class")
xyplot(wing ~ pop | elev, ylab = "Wing length", xlab = "Population", main =
"Elevation-specific relationship between wing and population")
```

10.3 ASIDE: USING SIMULATION TO ASSESS BIAS AND PRECISION OF AN ESTIMATOR

Let's quickly compare our parameter estimates with what we put into the data:

```
lm(wing ~ pop*elev)
all.effects

> lm(wing ~ pop*elev)

Call:
lm(formula = wing ~ pop * elev)

Coefficients:
(Intercept)        pop2        pop3        pop4        pop5       elev2
     38.859      -8.543      -4.793       4.702      13.410       6.437
       elev3   pop2:elev2  pop3:elev2  pop4:elev2  pop5:elev2  pop2:elev3
     11.213      -1.284       1.530       0.332       1.857       1.359
  pop3:elev3  pop4:elev3  pop5:elev3
       1.820       2.835      -6.609
> all.effects
 [1]  40  -10   -5    5   10    5   10   -2    3    0    4    4    0    3   -2
```

The coefficient estimates don't necessarily resemble very much the parameters from which we simulated these data; after all, our sample size is rather small. So, to reassure ourselves that these differences are simply due to sampling variation, we repeat this data generation-analysis cycle 1000 times and average over the random sampling variation to convince ourselves that the estimators from the linear model are indeed unbiased. Simulations of this kind can be done easily in R, and this is one of the great strengths of R.

```
n.iter <- 1000          # Desired number of iterations
estimates <- array(dim = c(n.iter, length(all.effects)))  # Data structure to
hold results

for(i in 1:n.iter) {    # Run simulation n.iter times
  print(i)              # Optional
  eps <- rnorm(n, 0, sigma)        # Residuals
```

```
y <- as.numeric(as.matrix(X) %*% as.matrix(all.effects) + eps) # Assemble data
fit.model <- lm(y ~ pop*elev)                        # Break down data
estimates[i,] <- fit.model$coefficients         # Save estimates of coefs.
}
```

Compare the input (i.e., the chosen effects) and the output when averaged over sampling variation:

```
print(apply(estimates, 2, mean), dig = 2)
all.effects

> print(apply(estimates, 2, mean), dig = 2)
 [1]  40.0102  -10.0452  -5.0717  4.9687  9.9524  5.0417  10.0419  -1.9761
 [9]   2.9493   -0.0226   4.0031  3.9407 -0.0039  3.0234  -1.9838
> all.effects
 [1]  40  -10  -5  5  10  5  10  -2  3  0  4  4  0  3  -2
```

These are much closer to our input parameters. Depending on the number of iterations, we can get arbitrarily close to the input. Alternatively, we could increase the sample size from 12 to 12000, and we would get estimates that are still closer to the input values.

10.4 ANALYSIS USING R

We continue with the analysis of our data set and use R to fit the main-effects model first.

```
mainfit <- lm(wing ~ elev + pop)
mainfit

> mainfit
[ ... ]
Coefficients:
(Intercept)       elev2        elev3          pop2
     38.736       6.924       11.095        -8.518
       pop3        pop4         pop5
     -3.676       5.757       11.826
```

Then, we fit the means parameterization of the interaction model.

```
intfit <- lm(wing ~ elev*pop-1-pop-elev)
intfit

> intfit <- lm(wing ~ elev*pop-1-pop-elev)
> intfit
[ ... ]
```

```
Coefficients:
elev1:pop1    elev2:pop1    elev3:pop1
      38.86         45.30         50.07
elev1:pop2    elev2:pop2    elev3:pop2
      30.32         35.47         42.89
elev1:pop3    elev2:pop3    elev3:pop3
      34.07         42.03         47.10
elev1:pop4    elev2:pop4    elev3:pop4
      43.56         50.33         57.61
elev1:pop5    elev2:pop5    elev3:pop5
      52.27         60.56         56.87
```

10.5 ANALYSIS USING WinBUGS

10.5.1 Main-Effects ANOVA Using WinBUGS

We fit the main-effects model in the effects parameterization because I find that easier to code. One minor feature in this analysis is the way in which we specify the priors for the elements of the parameter vectors: instead of looping over each of them, we now write them all out, since we have to set to zero the first (or another) level of each factor to make this fixed-effects model identifiable.

```
# Define model
sink("2w.anova.txt")
cat("
model {

# Priors
  alpha ~ dnorm(0, 0.001)              # Intercept
  beta.pop[1] <- 0                     # set to zero effect of 1st level
  beta.pop[2] ~ dnorm(0, 0.001)
  beta.pop[3] ~ dnorm(0, 0.001)
  beta.pop[4] ~ dnorm(0, 0.001)
  beta.pop[5] ~ dnorm(0, 0.001)
  beta.elev[1] <- 0                    # ditto
  beta.elev[2] ~ dnorm(0, 0.001)
  beta.elev[3] ~ dnorm(0, 0.001)
  sigma ~ dunif(0, 100)

# Likelihood
for (i in 1:n) {
   wing[i] ~ dnorm(mean[i], tau)
   mean[i] <- alpha + beta.pop[pop[i]] + beta.elev[elev[i]]
   }
```

```
# Derived quantities
tau <- 1 / ( sigma * sigma )
}
",fill=TRUE)
sink()
```

```
# Bundle data
win.data <- list(wing=wing, elev = as.numeric(elev), pop = as.numeric(pop), n =
length(wing))
```

```
# Inits function
inits <- function(){ list(alpha = rnorm(1), sigma = rlnorm(1) )}
```

```
# Parameters to estimate
params <- c("alpha", "beta.pop", "beta.elev", "sigma")
```

```
# MCMC settings
ni <- 1200
nb <- 200
nt <- 2
nc <- 3
```

```
# Start Gibbs sampling
out <- bugs(win.data, inits, params, "2w.anova.txt", n.thin=nt, n.chains=nc,
n.burnin=nb, n.iter=ni, debug = TRUE)
```

```
# Print estimates
print(out, dig = 3)
> print(out, dig = 3)
Inference for Bugs model at "2w.anova.txt", fit using WinBUGS,
 3 chains, each with 1200 iterations (first 200 discarded), n.thin = 2
 n.sims = 1500 iterations saved
```

	mean	sd	2.5%	25%	50%	75%	97.5%	Rhat	n.eff
alpha	38.701	1.216	36.355	37.870	38.760	39.510	40.945	1.006	680
beta.pop[2]	-8.447	1.476	-11.325	-9.492	-8.442	-7.428	-5.547	1.001	1500
beta.pop[3]	-3.567	1.458	-6.360	-4.512	-3.586	-2.664	-0.670	1.001	1500
beta.pop[4]	5.820	1.470	3.040	4.825	5.871	6.788	8.713	1.003	780
beta.pop[5]	11.841	1.454	9.090	10.850	11.860	12.770	14.725	1.001	1500
beta.elev[2]	6.901	1.156	4.714	6.106	6.923	7.671	9.184	1.004	560
beta.elev[3]	11.077	1.119	8.988	10.300	11.040	11.850	13.220	1.002	970
sigma	3.561	0.360	2.954	3.306	3.538	3.788	4.323	1.000	1500
deviance	320.983	4.308	314.900	317.900	320.400	323.200	331.600	1.001	1500

```
[ ... ]
DIC info (using the rule, pD = Dbar-Dhat)
pD = 7.8 and DIC = 328.8
DIC is an estimate of expected predictive error (lower deviance is better).
```

We get estimates that are fairly similar with the MLEs above. To see the estimate of the residual, you can type summary(mainfit).

10.5.2 Interaction-Effects ANOVA Using WinBUGS

We will specify the means parameterization for ease of coding and show how parameters in WinBUGS can be arrays with two (or more) dimensions. This is handy when organizing an analysis.

```
# Write model
sink("2w2.anova.txt")
cat("
model {

# Priors
  for (i in 1:n.pop){
    for(j in 1:n.elev) {
       group.mean[i,j] ~ dnorm(0, 0.0001)
    }
  }
sigma ~ dunif(0, 100)

# Likelihood
  for (i in 1:n) {
     wing[i] ~ dnorm(mean[i], tau)
     mean[i] <- group.mean[pop[i], elev[i]]
  }

# Derived quantities
  tau <- 1 / ( sigma * sigma )
}
",fill=TRUE)
sink()

# Bundle data
win.data <- list(wing=wing, elev = as.numeric(elev), pop = as.numeric(pop), n =
length(wing), n.elev = length(unique(elev)), n.pop = length(unique(pop)))

# Inits function
inits <- function(){list(sigma = rlnorm(1) )}

# Parameters to estimate
params <- c("group.mean", "sigma")

# MCMC settings
ni <- 1200
nb <- 200
nt <- 2
nc <- 3
```

```
# Start Gibbs sampling
out <- bugs(win.data, inits, params, "2w2.anova.txt", n.thin=nt, n.chains=nc,
n.burnin=nb, n.iter=ni, debug = TRUE)
# Print estimates
print(out, dig = 3)
> print(out, dig = 3)
Inference for Bugs model at "2w2.anova.txt", fit using WinBUGS,
 3 chains, each with 1200 iterations (first 200 discarded), n.thin = 2
 n.sims = 1500 iterations saved
```

	mean	sd	2.5%	25%	50%	75%	97.5%	Rhat	n.eff
group.mean[1,1]	38.864	1.644	35.500	37.790	38.885	39.982	42.040	1.001	1500
group.mean[1,2]	45.263	1.600	42.135	44.187	45.250	46.300	48.506	1.000	1500
group.mean[1,3]	50.062	1.645	46.784	48.987	50.080	51.110	53.330	1.000	1500
group.mean[2,1]	30.330	1.589	27.150	29.297	30.320	31.370	33.495	1.000	1500
group.mean[2,2]	35.405	1.637	32.284	34.340	35.390	36.532	38.495	1.004	720
group.mean[2,3]	42.905	1.564	39.708	41.865	42.930	43.950	45.850	1.003	780
group.mean[3,1]	34.071	1.634	30.975	32.977	34.065	35.120	37.346	1.002	1500
group.mean[3,2]	41.932	1.609	38.714	40.897	41.940	43.010	45.041	1.001	1500
group.mean[3,3]	47.077	1.628	43.855	46.017	47.040	48.160	50.290	1.004	500
group.mean[4,1]	43.508	1.663	40.345	42.370	43.440	44.580	46.895	1.001	1500
group.mean[4,2]	50.403	1.635	47.239	49.310	50.365	51.520	53.590	1.004	540
group.mean[4,3]	57.630	1.617	54.525	56.570	57.640	58.720	60.770	1.002	1500
group.mean[5,1]	52.246	1.608	49.054	51.177	52.260	53.350	55.411	1.002	1100
group.mean[5,2]	60.538	1.606	57.384	59.457	60.580	61.630	63.625	1.000	1500
group.mean[5,3]	56.893	1.621	53.670	55.877	56.870	57.962	60.095	1.002	1300
sigma	3.227	0.358	2.617	2.980	3.193	3.447	4.014	1.001	1500
deviance	309.326	6.734	299.100	304.500	308.350	313.000	324.952	1.001	1500

```
[...]
DIC info (using the rule, pD = Dbar-Dhat)
pD = 15.7 and DIC = 325.1
DIC is an estimate of expected predictive error (lower deviance is better).
```

We find the usual similarity between the Bayes and the maximum likelihood solution above (do `summary(intfit)`) and note in passing that the estimated number of parameters (pD) is pretty close to what we would expect it to be.

10.5.3 Forming Predictions

Let's present the Bayesian inference for the interaction-effects model in a graph showing the predicted response, analogous to least-square means in a classical analysis, for each combination of elevation and population (Fig. 10.3). This plot corresponds to the boxplot of the data set (Fig. 10.2); or selects the order of the predictions to match that in Fig. 10.2.

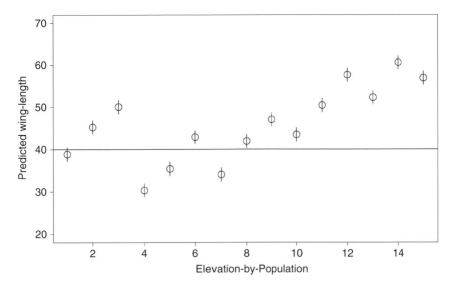

FIGURE 10.3 Predicted wing length of mourning cloaks for each elevation-population combination. Error bars are 1 SE.

```
or <- c(1,4,7,10,13,2,5,8,11,14,3,6,9,12,15)
plot(or, out$mean$group.mean, xlab = "Elev-by-Population", las = 1, ylab =
"Predicted wing-length", cex = 1.5, ylim = c(20, 70))
segments(or, out$mean$group.mean, or, out$mean$group.mean + out$sd$group.mean,
col = "black", lwd = 1)
segments(or, out$mean$group.mean, or, out$mean$group.mean - out$sd$group.mean,
col = "black", lwd = 1)
abline(h = 40)
```

10.6 SUMMARY

We have introduced the concepts of main and interaction effects and used R and WinBUGS to fit the corresponding two-way ANOVA models. In an aside, we have illustrated R's flexibility to conduct simulations to verify the effects of sampling variation on the parameter estimates.

EXERCISES

1. *Toy snake example*: Fit a two-way ANOVA with interaction to the toy example of Chapter 6 and see what happens to the nonidentifiable parameter.
2. *Swiss hare data*: Fit an ANOVA model to mean hare density to decide whether the effect of grassland and arable land use is the same in all regions. Regions and land use are somewhat confounded, but we ignore this here.

General Linear Model (ANCOVA)

11.1 INTRODUCTION

The "model of the mean," t-test, simple linear regression, and analysis of variance (ANOVA) are all just special cases of a very general and powerful statistical model, the general linear model (≠ GLM!; see Chapter 13). This model expresses a continuous response as a linear combination of the effects of discrete and/or continuous explanatory variables plus a single random contribution from a normal distribution, whose variance is estimated along with the coefficients of all discrete and continuous covariates and possible interactions.

Before this was widely recognized, people used to make a rather sharp and artificial distinction between linear models that contain categorical explanatory variables only and were called t-test or ANOVA models and those that contain continuous covariates only and were called regression models. Models that contained both types of explanatory variables were usually treated as ANOVAs with typically a single continuous covariate to correct for preexisting variation among experimental units. These

Introduction to WinBUGS for Ecologists
DOI: 10.1016/B978-0-12-378605-0.00011-9

FIGURE 11.1   Male asp viper (*Vipera aspis*), Switzerland, 2007. (*Photo T. Ott*)

models were called analysis of covariance (ANCOVA) models and that's why I use this term here.

Nowadays, in many practical applications, we typically have several explanatory variables of both types. In addition, we want to fit both main effects of these covariates and some or all their pairwise or even higher-order interactions. As an example, we here consider an ANCOVA model with an interaction between a discrete and a continuous covariate. We saw how to fit interactions between two discrete covariates in Chapter 10. Interactions between two continuous covariates are easy to fit: simply fit one additional covariate whose values are obtained by multiplication of the two main covariates.

In this chapter, we consider the relationship between body mass and body length of the asp viper (Fig. 11.1) in three populations: Pyrenees, Massif Central, and the Jura mountains. We are particularly interested in population-specific differences of the mass–length relationship, i.e., in the interaction between length and population. The means parameterization of the model we will fit can be written as (see Section 6.3.6)

$$y_i = \alpha_{j(i)} + \beta_{j(i)} * x_i + \varepsilon_i$$

and

$$\varepsilon_i \sim \text{Normal}(0, \sigma^2),$$

where y_i is the body mass of individual i, $\alpha_{j(i)}$ and $\beta_{j(i)}$ are the intercept and the slope, respectively, of the mass–length relationship in population j, x_i is the body length of snake i, and as usual, ε_i describes the combined effects of all unmeasured influences on the body mass of snake i and is assumed to behave like a normal random variable whose variance σ^2 we estimate.

The effects parameterization of the same model is this:

$$y_i = \alpha_{\text{Pyr}} + \beta_1 * x_{\text{MC}(i)} + \beta_2 * x_{\text{Jura}(i)} + \beta_3 * x_{\text{body}(i)} + \beta_4 * x_{\text{body}(i)} * x_{\text{MC}(i)}$$
$$+ \beta_5 * x_{\text{body}(i)} * x_{\text{Jura}(i)} + \varepsilon_i$$

In addition to y_i and ε_i that are as before, α_{Pyr} is the expected mass of snakes in the Pyrenees, β_1 is the difference between the expected mass of snakes in the Massif Central and that in the Pyrenees, and $x_{\text{MC}(i)}$ is the indicator for snakes caught in the Massif Central. β_2 is the difference between the expected mass in the Jura and that in the Pyrenees, $x_{\text{Jura}(i)}$ is the indicator for snakes in the Jura, β_3 is the slope of the regression of body mass on body length x_{body} in the Pyrenees, β_4 is the difference in that slope between the Massif Central and the Pyrenees, and β_5 is the difference of slopes between the Jura and the Pyrenees. Thus, snakes in the Pyrenees act as baseline with which snakes from the Massif Central and the Jura are compared, but as usual, this choice has no effect on inference.

11.2 DATA GENERATION

As always, we assume a balanced design simply for convenience of data generation.

```
n.groups <- 3
n.sample <- 10
n <- n.groups * n.sample # Total number of data points
x <- rep(1:n.groups, rep(n.sample, n.groups)) # Indicator for
population
pop <- factor(x, labels = c("Pyrenees", "Massif Central", "Jura"))
length <- runif(n, 45, 70)    # Obs. body length (cm) is rarely less
than 45
```

We build the design matrix of an interactive combination of length and population, inspect that, and select the parameter values, i.e., choose values for α_{Pyr}, β_1, β_2, β_3, β_4, and β_5.

```
Xmat <- model.matrix(~ pop*length)
print(Xmat, dig = 2)
beta.vec <- c(-250, 150, 200, 6, -3, -4)
```

Next, we build up the body mass measurements y_i by adding the residual to the value of the linear predictor, with residuals drawn from an appropriate zero-mean normal distribution. The value of the linear predictor is obtained by matrix multiplication of the design matrix (Xmat) and the parameter vector (beta.vec). Our vipers are probably all too fat, but that doesn't really matter for our purposes.

```
lin.pred <- Xmat[,] %*% beta.vec      # Value of lin.predictor
eps <- rnorm(n = n, mean = 0, sd = 10)  # residuals
mass <- lin.pred + eps                # response = lin.pred + residual
hist(mass)                            # Inspect what we've created
matplot(cbind(length[1:10], length[11:20], length[21:30]), cbind(mass[1:10],
mass[11:20], mass[21:30]), ylim = c(0, max(mass)), ylab = "Body mass (g)", xlab =
"Body length (cm)", col = c("Red","Green","Blue"), pch = c("P","M","J"), las = 1,
cex = 1.2, cex.lab = 1.5)
```

We have created a data set in which vipers from the Pyrenees have the steepest slope between mass and length, followed by those from the Massif Central and finally those in the Jura mountains (Fig. 11.2). Now let's disassemble these data.

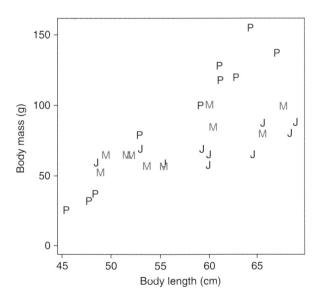

FIGURE 11.2   Simulated data set showing body mass versus length of 10 asp vipers in each of three populations (P – Pyrenees, M – Massif Central, and J – Jura).

11.3 ANALYSIS USING R

The code for an analysis in R is very parsimonious indeed:

```
summary(lm(mass ~ pop * length))
> summary(lm(mass ~ pop * length))
[...]
Coefficients:
```

	Estimate	Std. Error	t value	Pr(>|t|)	
(Intercept)	−246.6623	24.8588	−9.923	5.72e-10	***
popMassif Central	197.5543	37.8767	5.216	2.41e-05	***
popJura	241.5563	39.1405	6.172	2.24e-06	***
length	5.9623	0.4323	13.792	6.66e-13	***
popMassif Central:length	−3.8041	0.6631	−5.737	6.52e-06	***
popJura:length	−4.7114	0.6595	−7.144	2.20e-07	***

```
[...]
Residual standard error: 10.01 on 24 degrees of freedom
Multiple R-squared: 0.9114,    Adjusted R-squared: 0.8929
F-statistic: 49.37 on 5 and 24 DF, p-value: 7.48e-12
```

These coefficients can directly be compared with the beta vector since we simulated the data exactly in the default effects format of a linear model specified in R. The residual standard deviation is called residual standard error by R.

```
beta.vec
cat("And the residual SD was 10 \n")

> beta.vec
[1] −250  150  200    6   −3   −4
> cat("And the residual SD was 10 \n")
And the residual SD was 10
```

11.4 ANALYSIS USING WinBUGS
(AND A CAUTIONARY TALE
ABOUT THE IMPORTANCE OF
COVARIATE STANDARDIZATION)

In WinBUGS, I find it much easier to fit the means parameterization of the model, i.e., to specify three separate linear regressions for each mountain range. The effects (i.e., differences of intercept or slopes with reference to the Pyrenees) are trivially easy to recover as derived parameters by just adding a few WinBUGS code lines. This allows for better comparison between input and output values.

```
# Define model
sink("lm.txt")
cat("
model {

# Priors
 for (i in 1:n.group){
    alpha[i] ~ dnorm(0, 0.001)      # Intercepts
    beta[i] ~ dnorm(0, 0.001)       # Slopes
 }
 sigma ~ dunif(0, 100)           # Residual standard deviation
 tau <- 1 / ( sigma * sigma)

# Likelihood
 for (i in 1:n) {
    mass[i] ~ dnorm(mu[i], tau)
    mu[i] <- alpha[pop[i]] + beta[pop[i]]* length[i]
 }

# Derived quantities
# Define effects relative to baseline level
 a.effe2 <- alpha[2] - alpha[1]      # Intercept Massif Central vs. Pyr.
 a.effe3 <- alpha[3] - alpha[1]      # Intercept Jura vs. Pyr.
 b.effe2 <- beta[2] - beta[1]        # Slope Massif Central vs. Pyr.
 b.effe3 <- beta[3] - beta[1]        # Slope Jura vs. Pyr.

# Custom tests
 test1 <- beta[3] - beta[2]          # Slope Jura vs. Massif Central
}
",fill=TRUE)
sink()

# Bundle data
win.data <- list(mass = as.numeric(mass), pop = as.numeric(pop), length = length,
n.group = max(as.numeric(pop)), n = n)

# Inits function
inits <- function(){ list(alpha = rnorm(n.group, 0, 2), beta = rnorm(n.groups, 1,
1), sigma = rlnorm(1))}

# Parameters to estimate
parameters <- c("alpha", "beta", "sigma", "a.effe2", "a.effe3", "b.effe2",
"b.effe3", "test1")

# MCMC settings
ni <- 1200
nb <- 200
nt <- 2
nc <- 3
```

```
# Start Markov chains
out <- bugs(win.data, inits, parameters, "lm.txt", n.thin=nt, n.chains=nc,
n.burnin=nb, n.iter=ni, debug = TRUE)
```

This is a simple model that converges rapidly. We inspect the results and compare them with the "truth" in the data-generating random process as well as with the inference from R …

```
print(out, dig = 3)            # Bayesian analysis
beta.vec                       # Truth in the data-generating process
summary(lm(mass ~ pop * length))  # The ML solution again
```

… and are perplexed! WinBUGS claims that the Markov chains have converged (see Rhat values), but we get totally different estimates from what we should! Remember that alpha[1] and beta[1] in WinBUGS correspond to the intercept and the length main effect in the analysis in R and a.effe2, a.effe3. b.effe2, b.effe3 to the remaining terms of the analysis in R.

```
> print(out, dig = 3)# Bayesian analysis
Inference for Bugs model at "lm.txt", fit using WinBUGS,
 3 chains, each with 1200 iterations (first 200 discarded), n.thin = 2
 n.sims = 1500 iterations saved
```

	mean	sd	2.5%	25%	50%	75%	97.5%	Rhat	n.eff
alpha[1]	−100.524	31.918	−162.910	−122.400	−100.600	−79.105	−38.406	1.002	1500
alpha[2]	−16.420	26.238	−65.813	−34.060	−16.670	1.595	34.926	1.000	1500
alpha[3]	−2.613	26.127	−51.847	−20.862	−2.920	15.893	47.693	1.001	1500
beta[1]	3.442	0.557	2.336	3.066	3.443	3.813	4.540	1.002	1500
beta[2]	1.586	0.464	0.669	1.272	1.588	1.896	2.467	1.000	1500
beta[3]	1.205	0.438	0.367	0.885	1.207	1.503	2.059	1.002	1500
sigma	15.841	3.089	10.669	13.650	15.525	17.637	22.610	1.001	1500
a.effe2	84.103	38.764	6.313	57.752	84.620	110.300	160.082	1.002	1500
a.effe3	97.911	41.127	16.015	70.735	97.855	126.200	175.105	1.001	1500
b.effe2	−1.856	0.684	−3.178	−2.327	−1.856	−1.401	−0.479	1.002	1500
b.effe3	−2.237	0.708	−3.574	−2.716	−2.253	−1.777	−0.820	1.001	1500
test1	−0.381	0.633	−1.608	−0.782	−0.396	0.032	0.866	1.001	1500
deviance	249.071	8.258	232.847	243.300	249.000	254.800	264.552	1.000	1500

```
[ ... ]

> beta.vec                    # Truth in the data-generating process
[1]  −250   150   200    6   −3   −4
```

So lm() is able to recover the right parameter values, up to sampling and estimation error, but WinBUGS is not. Why is this?

The problem turns out to reside in the lack of standardization of the covariate length. In WinBUGS, it is always advantageous to scale covariates so that their extremes are not too far away from zero; otherwise, there may be nonconvergence or other problems. And the ugly thing here is that from looking at the convergence diagnostics (Rhat), we would never have guessed that there was a problem!

This example illustrates how useful it is to check the consistency of one's inference from WinBUGS with other sources, e.g., estimates from a simpler, but similar model run in WinBUGS or maximum likelihood estimates from another software. Know thy model! Alternatively, we could also have plotted the estimated regression lines into the observed data and would have seen easily that something was wrong.

As a quick check that lack of standardization was indeed the problem, we repeat both the maximum likelihood and the Bayesian analysis using a normalized version of the length covariate. This is simple; we just need to redefine the length covariate in the data list we pass to WinBUGS and can rerun the same code as before.

```
# Data passed to WinBUGS
win.data <- list(mass = as.numeric(mass), pop = as.numeric(pop), length =
as.numeric(scale(length)), n.group = max(as.numeric(pop)), n = n)

# Start Markov chains
out <- bugs(win.data, inits, parameters, "lm.txt", n.thin=nt, n.chains=nc,
n.burnin=nb, n.iter=ni, debug = FALSE)

...

# Inspect results
print(out, dig = 3)
> print(out, dig = 3)
[ ... ]
```

	mean	sd	2.5%	25%	50%	75%	97.5%	Rhat	n.eff
alpha[1]	97.796	3.406	90.679	95.820	97.905	100.100	104.252	1.004	1500
alpha[2]	74.952	3.504	67.864	72.690	75.030	77.292	81.580	1.004	1200
alpha[3]	66.535	3.653	59.009	64.060	66.660	68.995	73.265	1.007	350
beta[1]	41.243	3.158	34.829	39.150	41.215	43.332	47.316	1.005	870
beta[2]	14.486	3.763	7.318	11.947	14.570	16.920	22.112	1.002	1500
beta[3]	8.868	3.820	1.635	6.401	8.803	11.322	16.816	1.000	1500
sigma	10.587	1.678	7.937	9.380	10.370	11.510	14.516	1.005	630
a.effe2	−22.844	4.756	−31.771	−25.962	−22.815	−19.718	−13.498	1.003	1200
a.effe3	−31.262	5.000	−41.826	−34.470	−31.260	−28.050	−21.384	1.005	400
b.effe2	−26.757	4.802	−36.065	−30.032	−26.800	−23.620	−16.660	1.002	1500
b.effe3	−32.375	4.907	−41.445	−35.853	−32.350	−29.345	−22.162	1.001	1500
test1	−5.618	5.334	−16.024	−9.007	−5.836	−2.101	5.108	1.001	1500
deviance	225.218	4.558	218.700	221.900	224.500	227.800	235.500	1.009	270

```
[ ... ]
```

Compare with MLEs (R output slightly edited):

```
print(lm(mass ~ pop * as.numeric(scale(length)))$coefficients, dig = 4)

...

> print(lm(mass ~ pop * as.numeric(scale(length)))$coefficients, dig = 4)
                   (Intercept)             popMassif Central
                      98.94                    -22.95
                     popJura              as.numeric(scale(length))
                     -31.54                    41.78
popMassif Central:as.numeric(scale(length))  popJura:as.numeric(scale(length))
                     -26.66                    -33.01
```

Indeed, we now get consistent estimates in both analyses. Hence, previously the scale of the covariate didn't allow WinBUGS to converge, even though the Rhat values reported did indicate convergence. As a cautionary principle, we might therefore always consider to transform all covariates for WinBUGS, even if that slightly complicates presentation of results afterwards (for instance, in graphics). Transforming can mean centering, i.e., subtracting the mean, which changes the intercept only, but not the slope. Transforming can also mean normalizing, i.e., subtracting the mean and dividing the result by the standard deviation of the original covariate values. This changes both the intercept and the slope relative to an analysis with the original covariate. In the above case, centering will also work (want to try this out?).

11.5 SUMMARY

In a key chapter for your understanding of the modeling of grouped data, we have looked at the general linear model, or ANCOVA, in R and WinBUGS. We have focused on a model with one discrete and one continuous predictor and their interaction. Understanding ANCOVA is an important intermediate step to understanding the linear-mixed model in the next chapter.

EXERCISES

1. *Probability of a parameter*: What is the probability that the slope of the mass–length relationship of asp vipers is inferior in the Jura than in the Massif Central? Produce a graphical and a numerical answer.
2. *Related models*: Adapt the code to fit two variations of the model:
   - Fit different intercepts but a common slope
   - Fit the same intercept and the same slope

3. *Quadratic effects*: Add a quadratic term to the mass–length relationship, i.e., fit the model `pop + (length + length^2)`. You do not need to reassemble a data set that contains an effect of length squared, but you can simply take the data set we have already created in this chapter.

4. *Swiss hare data*: Fit an ANCOVA (`pop * year`, with year as a continuous explanatory variable) to the mean density. Also compute residuals and plot them to check for outliers.

CHAPTER

12

Linear Mixed-Effects Model

OUTLINE

12.1 INTRODUCTION

Mixed-effects or mixed models contain factors, or more generally covariates, with both fixed and random effects. During the last 15 years or so, the use of mixed models has greatly increased in statistical applications in ecology and related disciplines (Pinheiro and Bates, 2000; McCulloch and Searle, 2001; Lee et al., 2006; Littell et al., 2008). As explained in Chapter 9, there may be at least three benefits to assuming

FIGURE 12.1   Gravid female asp viper (*Vipera aspis*), France, 2008. (*Photo T. Ott*)

a set of parameters constitutes a random sample from some distribution, whose hyperparameters are then estimated as the main structural parameters of a model: increased scope of the inference, more honest accounting for system uncertainty, and efficiency of estimation.

In Chapter 9, we met a one-way analysis of variance model that, apart from an overall mean, contained only random effects and could be called a variance-components model. It could also be called a mixed model if the overall mean, the intercept, is viewed as a fixed effect, but this terminology is not standard. Here, we consider a classic mixed model that arises as a direct generalization of the analysis of covariance (ANCOVA) model in Chapter 11. We modify our asp viper (Fig. 12.1) data set from there just a bit and assume we now have measurements from a much larger number of populations, say, 56. A random-effects factor need not possess that many levels (some statisticians even fit a two-level factor such as sex as random; see Gelman, 2005), but one rarely sees fewer than, say, 5–10 or so parameters fitted as random effects. Estimating a variance with so few values, which are moreover unobserved, will not result in very precise and perhaps biased estimates (see also Lambert et al., 2005).

We resimulate some asp viper data using R code fairly similar to that in the previous chapter. However, we now constrain the values for at least one set of effects (intercepts and/or slopes) to come from a normal distribution: this is what the random-effects assumption means. There are at least three sets of assumptions that one may make about the random effects for the intercept and/or the slope of regression lines that are fitted to grouped (here, population-specific) data:

1. Only intercepts are random, but slopes are identical for all groups.
2. Both intercepts and slopes are random, but they are independent.
3. Both intercepts and slopes are random, and there is a correlation between them.

(An additional case, where slopes are random and intercepts are fixed, is not a sensible model in most circumstances.) Model No. 1 is often called a random-intercepts model, and both models No. 2 and 3 are also called random-coefficients models. Model No. 3 is the default in R's function lmer() in package lme4 when fitting a random-coefficients model.

We now first generate a random-coefficients data set under model No. 2, where both intercepts and slopes are uncorrelated random effects. We then fit both a random-intercepts (No. 1) and a random-coefficients model without correlation (No. 2) to these data (see Sections 12.2–12.4). Then, we generate a second data set that includes a correlation between random intercepts and random slopes and adopt the random-coefficients model with correlation between intercepts and slopes (No. 3) to analyze it (see Section 12.5).

This is a key chapter for your understanding of mixed models, and I expect its contents to be helpful for the general understanding of mixed models to many ecologists. A close examination of how such data can be assembled (i.e., simulated) will be an invaluable help for understanding how analogous data sets are broken down (i.e., analyzed) using mixed models. Indeed, I believe that very few strategies can be more effective to understand this type of mixed model than the combination of simulating data sets and describing the models fitted in WinBUGS syntax.

Here is one way in which to write the random-coefficients model without correlation between the random effects for mass y_i of snake i in population j:

$$y_i = \alpha_{j(i)} + \beta_{j(i)} * x_i + \varepsilon_i$$
$$\alpha_j \sim \text{Normal}(\mu_\alpha, \sigma_\alpha^2) \qquad \text{\# Random effects for intercepts}$$
$$\beta_j \sim \text{Normal}(\mu_\beta, \sigma_\beta^2) \qquad \text{\# Random effects for slopes}$$
$$\varepsilon_i \sim \text{Normal}(0, \sigma^2) \qquad \text{\# Residual ``random'' effects}$$

Exactly as in the ANCOVA model in Chapter 11, mass y_i is related to body length x_i of snake i by a straight-line relationship with population-specific values for intercept α_j and slope β_j. (These regression parameters vary by individual i according to their membership to population j.) However, both α_j and β_j are now assumed to come from an independent normal distribution, with means μ_α and μ_β and variances of σ_α^2 and σ_β^2, respectively.

The residuals ε_i for snake i are assumed to come from another independent normal distribution with variance σ^2.

12.2 DATA GENERATION

As always, we assume a balanced design for simple convenience, though that is not required to conduct mixed model analyses using restricted maximum likelihood (REML) in R or a Bayesian analysis in WinBUGS. Indeed, the flexibility with unbalanced data sets was one of the main reasons why REML-based mixed model estimation became so much more popular than estimation based on sums of squares decompositions (Littell et al., 2008).

```
n.groups <- 56              # Number of populations
n.sample <- 10              # Number of vipers in each pop
n <- n.groups * n.sample        # Total number of data points
pop <- gl(n = n.groups, k = n.sample)  # Indicator for population
```

We directly normalize covariate length to avoid trouble with WinBUGS.

```
# Body length (cm)
original.length <- runif(n, 45, 70)
mn <- mean(original.length)
sd <- sd(original.length)
cat("Mean and sd used to normalise.original length:", mn, sd, "\n\n")
length <- (original.length − mn) / sd
hist(length, col = "grey")
```

We build a design matrix without intercept.

```
Xmat <- model.matrix(~pop*length−1−length)
print(Xmat[1:21,], dig = 2)             # Print top 21 rows
```

Next, we choose parameter values, but this time, we need to constrain them, i.e., both values for the intercepts and those for the slopes will be drawn from two normal distributions for whom we specify four hyperparameters, i.e., two means (corresponding to μ_α and μ_β) and two standard deviations (SDs) (corresponding to the square root of σ_α^2 and σ_β^2). As residual variation, we use a mean-zero normal distribution with SD of 30.

```
intercept.mean <- 230       # mu_alpha
intercept.sd <- 20          # sigma_alpha
slope.mean <- 60            # mu_beta
slope.sd <- 30              # sigma_beta

intercept.effects<-rnorm(n = n.groups, mean = intercept.mean, sd = intercept.sd)
slope.effects <- rnorm(n = n.groups, mean = slope.mean, sd = slope.sd)
all.effects <- c(intercept.effects, slope.effects)   # Put them all together
```

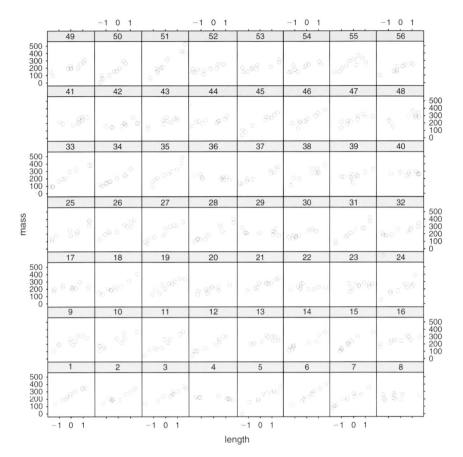

FIGURE 12.2   Trellis plot of the mass–length relationships in 56 asp viper populations (length has been standardized).

We assemble the measurements y_i as before.

```
lin.pred <- Xmat[,] %*% all.effects      # Value of lin.predictor
eps <- rnorm(n = n, mean = 0, sd = 30)   # residuals
mass <- lin.pred + eps                   # response = lin.pred + residual

hist(mass, col = "grey")                 # Inspect what we've created
```

We produce a trellis graph of the relationships in all ngroup populations (Fig. 12.2). Depending on the particular realization of the simulated stochastic system, we generally have quite a few fatties and may even have a few negative masses, but this doesn't really matter for our analysis.

```
library("lattice")
xyplot(mass ~ length | pop)
```

We can detect straight-line relationships between mass and length that differ among the 56 populations. What we can't see is the random-effects assumption built into the data set. That is, we are unable to distinguish a simple ANCOVA data set as in Chapter 11 from a mixed model data set as in this chapter.

12.3 ANALYSIS UNDER A RANDOM-INTERCEPTS MODEL

12.3.1 REML Analysis Using R

We first assume that the slope of the mass–length relationship is identical in all populations and that only the intercepts differ randomly from one population to another.

```
library('lme4')
lme.fit1 <- lmer(mass ~ length + (1 | pop), REML = TRUE)
lme.fit1
> lme.fit1
Linear mixed model fit by REML
Formula: mass ~ length + (1 | pop)
  AIC   BIC  logLik  deviance  REMLdev
  5873  5890  -2932    5872      5865
Random effects:
 Groups    Name       Variance   Std.Dev.
 pop       (Intercept) 260.60     16.143
 Residual              1930.94    43.942
Number of obs: 560, groups: pop, 56

Fixed effects:
              Estimate  Std. Error  t value
(Intercept)   226.527     2.846     79.59
length         59.647     1.916     31.13

Correlation of Fixed Effects:
       (Intr)
length 0.000
```

12.3.2 Bayesian Analysis Using WinBUGS

In our Bayesian analysis of the random-intercepts model, we use a suitably wide uniform distribution as a prior for the standard deviation of the random-effects distribution (Gelman, 2006).

```
# Write model
sink("lme.model1.txt")
cat("
model {

# Priors
  for (i in 1:ngroups){
     alpha[i] ~ dnorm(mu.int, tau.int)      # Random intercepts
  }

  mu.int ~ dnorm(0, 0.001)     # Mean hyperparameter for random intercepts
  tau.int <- 1 / (sigma.int * sigma.int)
  sigma.int ~ dunif(0, 100)    # SD hyperparameter for random intercepts

  beta ~ dnorm(0, 0.001)       # Common slope
  tau <- 1 / ( sigma * sigma)  # Residual precision
  sigma ~ dunif(0, 100)        # Residual standard deviation

# Likelihood
  for (i in 1:n) {
     mass[i] ~ dnorm(mu[i], tau)               # The random variable
     mu[i] <- alpha[pop[i]] + beta* length[i]        # Expectation
  }
}
",fill=TRUE)
sink()

# Bundle data
win.data <- list(mass = as.numeric(mass), pop = as.numeric(pop), length = length,
ngroups = max(as.numeric(pop)), n = n)

# Inits function
inits <- function(){list(alpha = rnorm(n.groups, 0, 2), beta = rnorm(1, 1, 1),
mu.int = rnorm(1, 0, 1), sigma.int = rlnorm(1), sigma = rlnorm(1))}

# Parameters to estimate
parameters <- c("alpha", "beta", "mu.int", "sigma.int", "sigma")

# MCMC settings
ni <- 2000
nb <- 500
nt <- 2
nc <- 3

# Start Gibbs sampling
out <- bugs(win.data, inits, parameters, "lme.model1.txt", n.thin=nt, n.chains=nc,
n.burnin=nb, n.iter=ni, debug = TRUE)
```

```
# Inspect results
print(out, dig = 3)
> print(out, dig = 3)
Inference for Bugs model at "lme.model.txt", fit using WinBUGS,
 3 chains, each with 2000 iterations (first 500 discarded), n.thin = 2
 n.sims = 2250 iterations saved
                mean     sd    2.5%     25%     50%     75%   97.5%  Rhat n.eff
[...]
beta          59.348  1.948  55.457  58.030  59.33  60.700  62.978 1.002  1300
mu.int       224.606  2.925 218.600 222.700 224.70 226.600 230.077 1.001  2200
sigma.int     16.507  2.760  11.582  14.580  16.42  18.230  22.356 1.012   180
sigma         44.120  1.409  41.570  43.150  44.08  45.040  46.958 1.002  1600
[...]

# Compare with input values
intercept.mean ; slope.mean ; intercept.sd ; slope.sd ; sd(eps)
> intercept.mean ; slope.mean ; intercept.sd ; slope.sd ; sd(eps)
[1] 230
[1] 60
[1] 20
[1] 30
[1] 29.86372
```

As usual with vague priors, the two analyses yield rather comparable results. Interestingly, the residual standard deviation in both is estimated too high. This is because we simulated the data to contain random variation among the slopes, but we did not fit this model, so this variation is unaccounted for and gets absorbed in the residual.

12.4 ANALYSIS UNDER A RANDOM-COEFFICIENTS MODEL WITHOUT CORRELATION BETWEEN INTERCEPT AND SLOPE

12.4.1 REML Analysis Using R

Next, we assume that both slopes and intercepts of the mass–length relationship differ among populations in the fashion of two independent random variables, i.e., we assume the absence of a correlation between intercept and slope. Thus, we will analyze the data under the same model that we used to generate our data set.

```
library('lme4')
lme.fit2 <- lmer(mass ~ length + (1 | pop) + ( 0+ length | pop))
lme.fit2

> lme.fit2
Linear mixed model fit by REML
```

```
Formula: mass ~ length + (1 | pop) + (0 + length | pop)
  AIC   BIC   logLik   deviance   REMLdev
 5598  5619   −2794      5596       5588
Random effects:
     Groups      Name          Variance    Std.Dev.
     pop         (Intercept)    274.37      16.564
     pop         length         1012.49     31.820
     Residual                   875.96      29.597
Number of obs: 560, groups: pop, 56

Fixed effects:
              Estimate   Std. Error   t value
(Intercept)   228.698       2.579      88.68
length         59.774       4.461      13.40

Correlation of Fixed Effects:
        (Intr)
length −0.002
```

12.4.2 Bayesian Analysis Using WinBUGS

Finally, here is the Bayesian analysis of the simple random-coefficients model:

```
# Define model
sink("lme.model2.txt")
cat("
model {

# Priors
  for (i in 1:ngroups){
      alpha[i] ~ dnorm(mu.int, tau.int)      # Random intercepts
      beta[i] ~ dnorm(mu.slope, tau.slope)   # Random slopes
  }

  mu.int ~ dnorm(0, 0.001)      # Mean hyperparameter for random intercepts
  tau.int <- 1 / (sigma.int * sigma.int)
  sigma.int ~ dunif(0, 100)     # SD hyperparameter for random intercepts

  mu.slope ~ dnorm(0, 0.001)    # Mean hyperparameter for random slopes
  tau.slope <- 1 / (sigma.slope * sigma.slope)
  sigma.slope ~ dunif(0, 100)   # SD hyperparameter for slopes

  tau <- 1 / ( sigma * sigma)   # Residual precision
  sigma ~ dunif(0, 100)         # Residual standard deviation

# Likelihood
  for (i in 1:n) {
```

```
    mass[i] ~ dnorm(mu[i], tau)
    mu[i] <- alpha[pop[i]] + beta[pop[i]]* length[i]
  }
}
",fill=TRUE)
sink()
```

```
# Bundle data
win.data < list(mass = as.numeric(mass), pop = as.numeric(pop), length = length,
ngroups = max(as.numeric(pop)), n = n)
```

```
# Inits function
inits <- function(){ list(alpha = rnorm(n.groups, 0, 2), beta = rnorm(n.groups,
10, 2), mu.int = rnorm(1, 0, 1), sigma.int = rlnorm(1), mu.slope = rnorm(1, 0, 1),
sigma.slope = rlnorm(1), sigma = rlnorm(1))}
```

```
# Parameters to estimate
parameters <- c("alpha", "beta", "mu.int", "sigma.int", "mu.slope", "sigma.
slope", "sigma")
```

```
# MCMC settings
ni <- 2000
nb <- 500
nt <- 2
nc <- 3
```

```
# Start Gibbs sampling
out <- bugs(win.data, inits, parameters, "lme.model2.txt", n.thin=nt, n.chains=nc,
n.burnin=nb, n.iter=ni, debug = TRUE)
```

This is still a relatively simple model that converges rapidly.

```
print(out, dig = 2)
> print(out, dig = 2)
Inference for Bugs model at "lme.model2.txt", fit using WinBUGS,
 3 chains, each with 2000 iterations (first 500 discarded), n.thin = 2
 n.sims = 2250 iterations saved
                mean    sd   2.5%    25%    50%    75%  97.5% Rhat n.eff
[...]
mu.int        227.22  2.68 221.90 225.50 227.20 229.00 232.40 1.00  1400
sigma.int      17.05  2.29  12.96  15.41  16.94  18.52  21.80 1.00  2200
mu.slope       58.49  4.57  49.48  55.37  58.49  61.59  66.96 1.00  2200
sigma.slope    32.48  3.39  26.50  30.16  32.24  34.63  39.74 1.00  2200
sigma          29.67  1.01  27.77  28.95  29.65  30.34  31.66 1.00  2200
 [...]
>
```

```
# Compare with input values
> intercept.mean  ;  slope.mean  ;  intercept.sd  ;  slope.sd  ;  sd(eps)
[1]  230
[1]  60
[1]  20
[1]  30
[1]  29.86372
```

The two sets of numbers agree rather nicely, as do the solutions obtained by `lmer()` and the input values. I emphasize again that using simulated data and successfully recovering the input values gives one the confidence that the analysis in WinBUGS has probably been specified correctly. For more complex models, this is helpful, since it's so easy to make mistakes!

Finally, we note that the realized values of the intercept and slope random effects are estimated and are returned by typing `ranef(lme.fit2)` for the analysis in R. They are also contained in the WinBUGS output that we get by typing `print(out, dig = 2)`.

12.5 THE RANDOM-COEFFICIENTS MODEL WITH CORRELATION BETWEEN INTERCEPT AND SLOPE

12.5.1 Introduction

The random-coefficients model with correlation is a simple extension of the previous model. The mass y_i of snake i in population j is assumed to be described by the following relations:

$$y_i = \alpha_{j(i)} + \beta_{j(i)} * x_i + \varepsilon_i$$

$$(\alpha_j, \beta_j) \sim \mathrm{MVN}(\mu, \Sigma) \qquad \text{# Bivariate normal random effects}$$

$$\mu = (\mu_\alpha, \mu_\beta) \qquad \text{# Mean vector}$$

$$\Sigma = \begin{pmatrix} \sigma_\alpha^2 & \sigma_{\alpha\beta} \\ \sigma_{\alpha\beta} & \sigma_\beta^2 \end{pmatrix} \qquad \text{# Variance–covariance matrix}$$

$$\varepsilon_i \sim \mathrm{Normal}(0, \sigma^2) \qquad \text{# Residual "random" effects}$$

As before, the mass y_i of snake i in population j is related to its body length x_i by a straight-line relationship with population-specific values for intercept α_j and slope β_j. But now, pairs of α_j and β_j from the same population are assumed to come from a multivariate normal distribution (actually, here, a bivariate normal) with mean vector μ and variance–covariance matrix Σ. The latter contains the variances of the intercept (σ_α^2) and the slope (σ_β^2) in the diagonal and the covariance between α_j and β_j ($\sigma_{\alpha\beta}$) in the off-diagonals. As before, the residuals ε_i for snake i are assumed to come from an independent (univariate) normal distribution with variance σ^2. The interpretation of the covariance is such that positive values indicate a steeper mass–length relationship for snakes with a greater mass.

12.5.2 Data Generation

We generate data under the random-coefficients model.

```
n.groups <- 56
n.sample <- 10
n <- n.groups * n.sample
pop <- gl(n = n.groups, k = n.sample)
```

We generate the covariate length.

```
original.length <- runif(n, 45, 70) # Body length (cm)
mn <- mean(original.length)
sd <- sd(original.length)
cat("Mean and sd used to normalise.original length:", mn, sd, "\n\n")
length <- (original.length - mn) / sd
hist(length, col = "grey")
```

We build the same design matrix as before.

```
Xmat <- model.matrix(~pop*length-1-length)
print(Xmat[1:21,], dig = 2)          # Print top 21 rows
```

We choose the parameter values, i.e., the population-specific intercepts and slopes from a bivariate normal distribution (available in the R package MASS) whose hyperparameters (two means and the four cells of the variance–covariance matrix) we need to specify. We use again as residual variation a mean-zero normal distribution with SD of 30.

```
library(MASS)                        # Load MASS
?mvrnorm                             # Calls help file

intercept.mean <- 230                # Values for five hyperparameters
intercept.sd <- 20
slope.mean <- 60
slope.sd <- 30
intercept.slope.covariance <- 10

mu.vector <- c(intercept.mean, slope.mean)
var.cova.matrix <- matrix(c(intercept.sd^2,intercept.slope.covariance,
intercept.slope.covariance, slope.sd^2),2,2)

effects <- mvrnorm(n = n.groups, mu = mu.vector, Sigma = var.cova.matrix)
effects                              # Look at what we've created
apply(effects, 2, mean)
var(effects)
```

```
intercept.effects <- effects[,1]
slope.effects <- effects[,2]
all.effects <- c(intercept.effects, slope.effects)     # Put them all together
```

Assemble the measurements y_i.

```
lin.pred <- Xmat[,] %*% all.effects          # Value of lin.predictor
eps <- rnorm(n = n, mean = 0, sd = 30)       # residuals
mass <- lin.pred + eps                       # response = lin.pred + residual

hist(mass, col = "grey")                     # Inspect what we've created
```

Again, negative masses are possible for some realizations of the data set. We look at the simulated data set:

```
library("lattice")
xyplot(mass ~ length | pop)
```

Now, we analyze this second data set allowing for a nonzero covariance between intercept and slope effects.

12.5.3 REML Analysis Using R

The model with an intercept–slope correlation is the default when specifying a random-coefficients model in R using function `lmer()`.

```
library('lme4')
lme.fit3 <- lmer(mass ~ length + (length | pop))
lme.fit3
> lme.fit3
Linear mixed model fit by REML
Formula: mass ~ length + (length | pop)
  AIC    BIC   logLik  deviance  REMLdev
 5624   5650   -2806     5620      5612
Random effects:
 Groups    Name         Variance  Std.Dev.  Corr
 pop       (Intercept)  255.86    15.996
           length       652.01    25.534    0.333
 Residual               979.29    31.294
Number of obs: 560, groups: pop, 56

Fixed effects:
             Estimate  Std. Error  t value
(Intercept)   233.342       2.554    91.38
length         68.811       3.698    18.61

Correlation of Fixed Effects:
       (Intr)
length 0.254
```

12.5.4 Bayesian Analysis Using WinBUGS

Here is one way in which to specify a Bayesian analysis of the random-coefficients model with correlation. For a different and more general way to allow for correlation among two or more sets of random effects in a model, see Gelman and Hill (2007, p. 376–377).

```
# Define model
sink("lme.model3.txt")
cat("
model {

# Priors
  for (i in 1:ngroups){
     alpha[i] <- B[i,1]
     beta[i] <- B[i,2]
     B[i,1:2] ~ dmnorm(B.hat[i,], Tau.B[,])
     B.hat[i,1] <- mu.int
     B.hat[i,2] <- mu.slope
}

  mu.int ~ dnorm(0, 0.001)      # Hyperpriors for random intercepts
  mu.slope ~ dnorm(0, 0.001)    # Hyperpriors for random slopes

  Tau.B[1:2,1:2] <- inverse(Sigma.B[,])
  Sigma.B[1,1] <- pow(sigma.int,2)
  sigma.int ~ dunif(0, 100)      # SD of intercepts
  Sigma.B[2,2] <- pow(sigma.slope,2)
  sigma.slope ~ dunif(0, 100)  # SD of slopes
  Sigma.B[1,2] <- rho*sigma.int*sigma.slope
  Sigma.B[2,1] <- Sigma.B[1,2]
  rho ~ dunif(-1,1)
  covariance <- Sigma.B[1,2]

  tau <- 1 / ( sigma * sigma)     # Residual
  sigma ~ dunif(0, 100)           # Residual standard deviation

# Likelihood
  for (i in 1:n) {
     mass[i] ~ dnorm(mu[i], tau)                  # The "residual" random variable
     mu[i] <- alpha[pop[i]] + beta[pop[i]]* length[i]    # Expectation
  }
}
",fill=TRUE)
sink()

# Bundle data
win.data <- list(mass = as.numeric(mass), pop = as.numeric(pop), length = length,
ngroups = max(as.numeric(pop)), n = n)
```

```
# Inits function
inits <- function(){ list(mu.int = rnorm(1, 0, 1), sigma.int = rlnorm(1), mu.slope
= rnorm(1, 0, 1), sigma.slope = rlnorm(1), rho = runif(1, -1, 1), sigma =
rlnorm(1))}

# Parameters to estimate
parameters <- c("alpha", "beta", "mu.int", "sigma.int", "mu.slope", "sigma.slope",
"rho", "covariance", "sigma")

# MCMC settings
ni <- 2000
nb <- 500
nt <- 2
nc <- 3

# Start Gibbs sampler
out <- bugs(win.data, inits, parameters, "lme.model3.txt", n.thin=nt, n.chains=nc,
n.burnin=nb, n.iter=ni, debug = TRUE)
```

We inspect the results and compare them with the frequentist analysis in Section 12.5.3 and find the usual comforting agreement between the two approaches (note rho in the Bayesian analysis has to be compared with Corr in the frequentist analysis).

```
print(out, dig = 2)          # Bayesian analysis
lme.fit3                     # Frequentist analysis
```

As usual, the approximate solution given by lmer() comes reasonably close to the exact solution from the Bayesian analysis (Gelman and Hill, 2007). Even though convergence is achieved fairly rapidly in the latter, it often takes much longer to obtain the exact Bayesian solution in a mixed model analysis. So there is a price to pay for enjoying the advantages of the Bayesian analysis, and this price can be fairly high when using Win-BUGS to fit more complex models.

For some realizations of the data set, the covariance may be estimated at a negative value, even though we've built the data with a positive covariance in the parent (statistical) population of intercepts and slopes. This is a reflection of both sampling variation and estimation error. Also look at how imprecise the estimate for the covariance is; covariances are even harder to estimate than variances. R doesn't return a standard error for that estimate.

12.6 SUMMARY

We have introduced the classic mixed ANCOVA model with random intercepts, random slopes, and the possibility of an intercept–slope covariance. Understanding the material presented in this chapter is essential for a thorough understanding of much of the current mixed modeling in ecology. The ideas presented here appear over and over again, in later chapters of this book, as well as in the applied work of many quantitative ecologists.

EXERCISES

1. *Specification of fixed- and random effects in WinBUGS*: The WinBUGS model description for the random-intercepts, random-slope model (i.e., the second one we fit in this chapter) is very similar to the fixed-effects "version" of the same model, i.e., the one we fitted in Chapter 11. Without looking at the WinBUGS model description in that chapter, take the linear mixed model description for WinBUGS from the current chapter and change it back to a fixed-effects ANCOVA with population-specific intercepts and slopes, i.e., corresponding to what you would fit in R as `lm(mass ~ pop*length)`.

2. *Swiss hare data*: Fit a random-coefficients regression without intercept–slope correlation to mean density (i.e., `~ population * year`, with year continuous).

Introduction to the Generalized Linear Model: Poisson "t-test"

13.1 INTRODUCTION

The unification of a large number of statistical methods such as regression, analysis of variance (ANOVA), and analysis of covariance (ANCOVA) under the umbrella of the *general linear model* was a big advancement for applied statistics. However, even more significant was the unification of an even wider range of statistical methods within the class of the *generalized linear model* or GLM in 1972 by Nelder and Wedderburn (also see McCullagh and Nelder, 1989). They showed that a large number of techniques previously thought of as representing quite separate types of analyses, including logistic regression, multinomial regression, Chi-square tests, log-linear models, as well as the general linear model, could all be represented as special cases of a generalized version of a linear model. In that way, much of what was well understood for the linear model could be carried over to that much larger class of models.

The two main ideas of the GLM are that, first, a *transformation of the expectation* of the response $E(y)$ is expressed as a linear combination of covariate effects rather than the expected (mean) response itself. And second, for the random part of the model, *distributions other than the normal* can be chosen, e.g., Poisson or binomial.

Formally, a GLM is described by the following three components:

1. a *statistical distribution* is used to describe the random variation in the response y; this is the stochastic part of the system description,
2. a so-called *link function* g, that is applied to the expectation of the response $E(y)$, and
3. a *linear predictor*, which is a linear combination of covariate effects that are thought to make up $g(E(y))$; this is the systematic or deterministic part of the system description.

Binomial, Poisson, and normal are probably the three most widely used statistical distributions in a GLM (see Chapter 6). The former two are distributions for non-negative, discrete responses and therefore suitable to describe counts. The normal is the most widely used distribution for continuous responses such as measurements. The three most widely used link functions are the identity, `logit(=log(odds)=log(x/(1-x)))`, and the `log`. For various reasons, one link function is typically advantageous, although not obligate, for each of these distributions. For instance, the normal distribution combined with an identity link yields the general linear model; the Poisson with a log link yields a log-linear model; and the binomial with a logit link yields a logistic regression. Hence, all the normal linear models seen in Chapters 4–11 are simply special cases of a GLM.

In the next nine chapters, we will go through a progression from simple to more complex models for Poisson and binomial responses. As for normal linear models, we begin again with what might be called a "Poisson t-test" in the sense that it consists of a comparison of two groups. To better see the analogy with the normal linear model, we start by writing the model for the normal two-group comparison (see Chapter 7) in GLM format:

1. Distribution: $y_i \sim \text{Normal}\,(\mu_i, \sigma^2)$
2. Link function: identity, i.e., $\mu_i = E(y_i) = \text{linear predictor}$
3. Linear predictor: $\alpha + \beta * x_i$

Next, we generalize this model to count data. The inferential situation considered is that of counts (C) of Brown hares (Fig. 13.1) in a sample of 10 arable and 10 grassland study areas. We wonder whether hare density depends on land-use.

FIGURE 13.1   Brown hare (*Lepus europaeus*), Germany, 2008. (*Photo N. Zbinden*)

The typical distribution assumed for such counts is a Poisson, which applies when counted things are distributed independently and randomly and samples of equal size are taken randomly. Then, the number of hares counted per study area (C) will be described by a Poisson. The Poisson has a single parameter, the expected count λ, that is often called the intensity and here represents the mean hare density. In contrast to the normal, the Poisson variance is not a free parameter but is equal to the mean λ. For a Poisson-distributed random variable C, we write $C \sim \text{Poisson}(\lambda)$.

If hare density depends on land-use, i.e., is different in arable and grassland areas, the assumption of a constant mean density across all 20 study areas is not realistic. And in a "Poisson t-test" we are specifically interested in whether hare density differs between grassland and arable areas. Therefore, here is a model for hare count C_i in area i:

1. Distribution: $C_i \sim \text{Poisson}(\lambda_i)$
2. Link function: log, i.e., $\log(\lambda_i) = \log(E(C_i)) = \text{linear predictor}$
3. Linear predictor: $\alpha + \beta * x_i$

In words, hare count C_i in area i is distributed as a Poisson random variable with mean $E(C_i) = \lambda_i$. The log-transformation of λ_i is assumed to be a linear function $\alpha + \beta * x_i$, where α and β are unknown constants and x_i is the value of an area-specific covariate. If x_i is an indicator for arable areas, then α becomes the mean hare density on a log-scale in grassland areas and β, again on a log-scale, is the difference in mean density between the two land-use types.

13.2 AN IMPORTANT BUT OFTEN FORGOTTEN ISSUE WITH COUNT DATA

Whenever we interpret λ as the mean *hare density*, we make the implicit assumption that *every individual hare is indeed seen*, i.e., that detection probability (p) is equal to 1. This is not very likely for hares nor indeed for any wild animal because typically some individuals are overlooked (Yoccoz et al., 2001; Kéry, 2002; Williams et al., 2002; Kéry and Juillerat, 2004; Schmidt, 2005, 2008). Alternatively, we may assume that the *proportion of hares overlooked per area is the same*, on average, in both land-use types. In that case, counts are considered just an index to absolute density, i.e., a measure for *relative density*, and what we model as the Poisson parameter λ_i is in reality the product between absolute hare density and the proportion p of hares seen. Only by making the assumption that p is identical, on average, in both land-use types may we validly interpret a mean *difference* between counts in arable and grassland areas as an indication of a difference in true hare density. See Chapters 20 and 21 for more on this important topic, the distinction between the imperfectly observed true state and the observed data, or, between the ecological and the observation processes underlying ecological field data.

13.3 DATA GENERATION

For now, we simulate and analyze hare counts under the assumption that detectability is perfect. First we need an indicator for land-use:

```
n.sites <- 10
x <- gl(n = 2, k = n.sites, labels = c("grassland", "arable"))
n <- 2*n.sites
```

Let the mean hare density in grassland and arable areas be 2 and 5 hares, respectively. Then, $\alpha = \log(2) = 0.69$ and $\log(5) = \alpha + \beta$, thus, $\beta = \log(5) - \log(2) = 0.92$. Therefore, the expected density λ_i is given by:

```
lambda <- exp(0.69 + 0.92*(as.numeric(x)-1)) # x has levels 1 and 2, not 0 and 1
```

We add the noise that comes from a Poisson distribution and inspect the hare counts we've thus generated (Fig. 13.2):

```
C <- rpois(n = n, lambda = lambda)              # Add Poisson noise
aggregate(C, by = list(x), FUN = mean)          # The observed means
boxplot(C ~ x, col = "grey", xlab = "Land-use", ylab = "Hare count", las = 1)
```

Again, we can get a feel for the strong effects of chance (sampling variation) by repeatedly generating hare counts and observing by how much they vary from one sample of 20 counts to another sample of 20 counts.

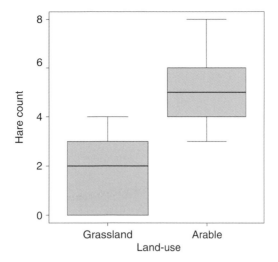

FIGURE 13.2   Relationship between hare count and land-use.

13.4 ANALYSIS USING R

We fit the "Poisson t-test" using the R function `glm(..., family= poisson)`. To test whether mean density in grassland differs from that in arable areas, we use the t-test provided by the function `summary()` or a likelihood ratio test from `anova()`. There is no big difference here in terms of the inferences.

```
poisson.t.test <- glm(C ~ x, family = poisson)    # Fit the model
summary(poisson.t.test)                           # t-Test
anova(poisson.t.test, test = "Chisq")             # Likelihood ratio test (LRT)
```

13.5 ANALYSIS USING WinBUGS

Let's now fit the "Poisson t-test" in WinBUGS. To do this, we will take the code from the normal t-test (Chapter 7) and adapt it to the Poisson GLM case. In addition, we will do two more things in the WinBUGS program below:

1. compute Pearson residuals to assess model fit and
2. conduct a posterior predictive check including a Bayesian p-value as we did for normal linear regression in Chapter 8 (and will do for a "generalized" Poisson regression in Chapter 21).

```
# Define model
sink("Poisson.t.test.txt")
cat("
model {

# Priors
 alpha ~ dnorm(0,0.001)
 beta ~ dnorm(0,0.001)

# Likelihood
 for (i in 1:n) {
    C[i] ~ dpois(lambda[i])
    log(lambda[i]) <- alpha + beta *x[i]

# Fit assessments
    Presi[i] <- (C[i] - lambda[i]) / sqrt(lambda[i])          # Pearson residuals
    C.new[i] ~ dpois(lambda[i])                               # Replicate data set
    Presi.new[i] <- (C.new[i] - lambda[i]) / sqrt(lambda[i])      # Pearson resi
    D[i] <- pow(Presi[i], 2)
    D.new[i] <- pow(Presi.new[i], 2)
 }

# Add up discrepancy measures
fit <- sum(D[])
fit.new <- sum(D.new[])
 }
",fill=TRUE)
sink()

# Bundle data
win.data <- list(C = C, x = as.numeric(x)-1, n = length(x))

# Inits function
inits <- function(){ list(alpha=rlnorm(1), beta=rlnorm(1))}

# Parameters to estimate
params <- c("lambda","alpha", "beta", "Presi", "fit", "fit.new")

# MCMC settings
nc <- 3
ni <- 3000
nb <- 1000
nt <- 2

# Start Gibbs sampler
out <- bugs(data=win.data, inits=inits, parameters.to.save=params,
model.file="Poisson.t.test.txt", n.thin=nt, n.chains=nc, n.burnin=nb, n.iter=ni,
debug = TRUE)
```

13.5.1 Check of Markov Chain Monte Carlo Convergence and Model Adequacy

The first two things to do before even looking at the estimates of a model fitted using MCMC should really be to check (1) that the Markov chains have converged and (2) that the fitted model is adequate for the data set. We do both here in an exemplary manner.

Convergence—Again, we can assess convergence by graphical means (typically directly within WinBUGS) or using a numerical summary, the Brooks–Gelman–Rubin statistic, which R2WinBUGS calls Rhat. Rhat is about 1 at convergence, with 1.1 often taken an acceptable threshold. We will look at the Rhat values first.

```
print(out, dig = 3)
```

If we briefly look at the second to last column in the table, we see that the chains for all parameters seem to have converged admirably. For larger models with many more parameters, a summary of this summary table may be useful. For instance, we may ask which (if any) parameters have a value of Rhat greater than 1.1. Or we can draw a histogram of the Rhat values.

```
which(out$summary[,8] > 1.1)      # which value in the 8th column is > 1.1 ?
> which(out$summary[,8] > 1.1)
named integer(0)                  # So here we have none

hist(out$summary[,8], col - "grey", main = "Rhat values")
```

So, as expected in this simple model fitted to a "good" data set, there is no problem with convergence.

Residuals and posterior predictive check—We do the analogous to what we did in the normal linear regression example in Chapter 8. That is, we plot the residuals first and then plot the two fit statistics (for the actual data set and for the perfect, new, data sets) against each other and compute the Bayesian p-value as a numerical summary of overall lack of fit. The fit statistic for the new data sets represents, in a way, the reference distribution for the chosen test statistic, here, the sum of squared Pearson residuals.

For GLMs other than the normal linear model, the variability of the response depends on the mean response. To get residuals with approximately constant variance, Pearson residuals are often computed. They are obtained by dividing the raw residuals $(y_i - \bar{y})$ by the standard deviation of y; see also WinBUGS code above.

```
plot(out$mean$Presi, las = 1)
abline(h = 0)
```

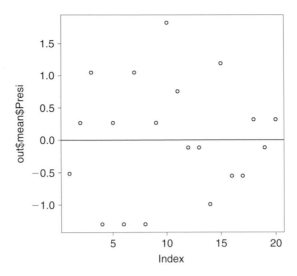

FIGURE 13.3   Pearson residuals for the hare counts.

There is no obvious sign of lack of fit for any particular data point (Fig. 13.3). Next, we conduct a posterior predictive check (Fig. 13.4):

```
plot(out$sims.list$fit, out$sims.list$fit.new, main = "Posterior predictive check
\nfor sum of squared Pearson residuals", xlab = "Discrepancy measure for actual data set",
ylab = "Discrepancy measure for perfect data sets")
abline(0,1, lwd = 2, col = "black")
```

Of course, this looks perfect and computation of the Bayesian p-value (below) confirms this impression. Here, we compute the Bayesian p-value outside WinBUGS in R. This is easier, but of course, we need to have saved the Markov chains for both fit and fit.new.

```
mean(out$sims.list$fit.new > out$sims.list$fit)
> mean(out$sims.list$fit.new > out$sims.list$fit)
[1] 0.624
```

13.5.2 Inference Under the Model

Now that we are convinced that the model is adequate for these data, we inspect the estimates and compare them with what we put into the data set, as well as what the frequentist analysis in R tells us.

```
print(out, dig = 3)
```

A comparison of the Bayesian solution with the input values that were used for generating the data set ($\alpha = 0.69$, $\beta = 0.92$) and the solution given by glm() shows a reasonably decent consistency (in view of the small sample size).

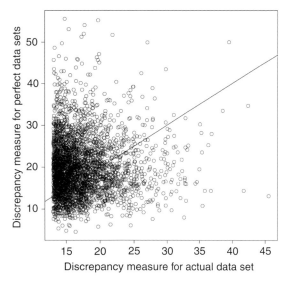

FIGURE 13.4 Graphical posterior predictive check based on the sum of squared Pearson residuals. The Bayesian p-value (0.62) is the proportion of points above the line. The hard boundary on the left is because of the fact that the discrepancy measure cannot be smaller than that corresponding to the maximum likelihood estimate.

```
summary(poisson.t.test)
```

So is there a difference in hare density according to land-use? Let's look at the posterior distribution of the coefficient for arable (Fig. 13.5).

```
hist(out$sims.list$beta, col = "grey", las = 1, xlab = "Coefficient for arable",
    main = "")
```

The posterior distribution does not overlap zero, so arable sites really do appear to have a different hare density than grassland sites. The same conclusion is arrived at when looking at the 95% credible interval of β in the summary of the analysis mentioned earlier: (0.64–1.72).

Finally, we will form predictions for presentation. Predictions are the expected values of the response under certain conditions, such as for particular covariate values. We have seen earlier that predictions are a valuable means for synthesizing the information that a model extracts from a data set. In a Bayesian analysis, forming predictions is easy. Predictions are just another form of unobservables, such as parameters or missing values. Therefore, we can base inference about predictions on their posterior distributions.

To summarize what we have learned about the differences in hare densities in grassland and arable study areas, we plot the posterior distributions of the expected hare counts (λ) for both habitat types (Fig. 13.6). We obtain the expected hare counts by exponentiating parameter α and $\alpha + \beta$, respectively.

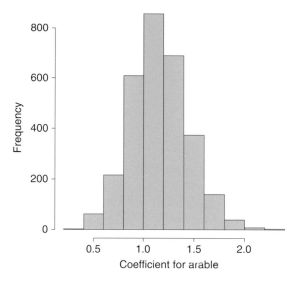

FIGURE 13.5   Posterior distribution of the coefficient of arable.

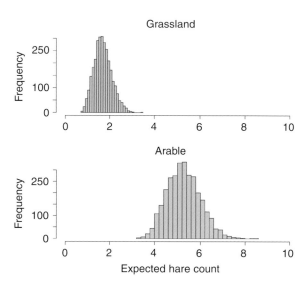

FIGURE 13.6   Posterior distribution of mean hare density in grassland (top) and arable areas (bottom).

```
par(mfrow = c(2,1))
hist(exp(out$sims.list$alpha), main = "Grassland study areas", col = "grey", xlab =
"", xlim = c(0,10), breaks = 20)
hist(exp(out$sims.list$alpha + out$sims.list$beta), main = "Arable study areas",
col = "grey", xlab = "Expected hare count", xlim = c(0,10), breaks = 20)
```

13.6 SUMMARY

We have introduced the generalized linear model, or GLM, where effects of covariates are linear in the transformed expectation of a response, which may come from a distribution other than the normal. The GLM is another key concept that appears over and over again in modern applied statistics in empirical sciences such as ecology. Therefore, we will deepen our understanding of this essential model class in subsequent chapters. Furthermore, we will combine the GLM and the mixed model to arrive at the most complex model considered in this book, the generalized linear mixed model, in Chapters 16 and 19–21.

EXERCISES

1. *Predictions*: Within the WinBUGS code, add a line that directly computes the mean hare density in arable areas.
2. *Derived quantities*: Summarize the posterior distribution for the difference in mean hare density in grassland and arable areas.
3. *Zeroes (migrating raptors)*: This fine example is borrowed from Bernardo (2003). It beautifully illustrates the power of Bayesian inference based on the posterior distribution of the unobservables (parameters, etc.). Ornithologists frequently count migrating birds of prey at places where they concentrate in spring or autumn, e.g., along coasts, mountain ridges, or on isthmuses. Assume that at a certain place, one rare raptor species had not been seen during 10 consecutive days. What is the probability that we see at least one on day 11? What is the probability that we see two or more? In your solution, make explicit your reasoning for using the particular statistical model you choose and discuss a few of its assumptions that may not hold in reality (e.g., serial independence, constancy of rates).
4. *Zeroes (contrast estimate)*: Assume that no hare was ever observed in grassland areas, i.e., that the counts in all 10 grassland areas were zero. Try to fit the Poisson t-test using R and using WinBUGS.
5. *Swiss hare data*: Compare mean counts (not density) in arable and in grassland areas in one selected year (e.g., 2000; take the smallest counts when there is more than one per year). When taking these counts as observations from an identical Poisson distribution, is there anything that strikes you as inadequate?

Overdispersion, Zero-Inflation, and Offsets in the GLM

Two features specific to nonnormal generalized linear models (GLMs) are overdispersion and offsets. Zero-inflation can be called a specific form of overdispersion: there are more zeroes than expected. Here, we briefly deal with each of them in the context of the hare counts example.

Introduction to WinBUGS for Ecologists
DOI: 10.1016/B978-0-12-378605-0.00014-4

14.1 OVERDISPERSION

14.1.1 Introduction

In both distributions commonly used to model counts (Poisson and binomial), the dispersion (the variability in the counts) is not a free parameter but instead is a function of the mean. The variance is equal to the mean (λ) for the Poisson and equal to the mean ($N * p$) times $1 - p$ for the binomial distribution (see Chapters 17–19). This means that *for a Poisson or binomial random variable, the models for the counts come with a "built-in" variability* and the magnitude of that variability is known. In an analysis of deviance conducted in a classical statistical analysis of the model, the residual deviance of the model will be about the same magnitude as the residual degrees of freedom, i.e., the mean deviance ratio (= residual deviance/residual df) is about 1.

However, in real life, count data are almost always more variable than expected under the Poisson or binomial models. This is called overdispersion or extra-Poisson or extra-binomial variation and means that the residual variation is larger than prescribed by a Poisson or binomial. Overdispersion can occur because there are hidden correlations that have not been included in the model, e.g., when individuals in family groups are assumed to be independent, or when important covariates have not been included. When overdispersion is not modeled, tests and confidence intervals will be overconfident (although means won't normally be biased). Therefore, overdispersion should be tested and corrected for when necessary.

The simplest way to correct for overdispersion in a classical analysis is by the quasi-likelihood (McCullagh and Nelder, 1989) and by using `family=quasipoisson` (or `quasibinomial`) in the R function `glm()`. Using WinBUGS, there are several ways in which one can account for overdispersion. One is to specify a distribution that is overdispersed relative to the Poisson, such as the negative binomial. Another solution, and the one we illustrate here, is to add into the linear predictor for the Poisson intensity a normally distributed random effect. Technically, this model is then a Poisson generalized linear mixed model (GLMM; see Chapter 16 for a more formal introduction). It is sometimes called a Poisson-lognormal model (Millar 2009).

14.1.2 Data Generation

We generate a slightly modified hare count data set, where in addition to the land-use difference in mean density, there is also a normally distributed site-specific effect in the linear predictor. For illustrative purposes, we also generate a sister data set without overdispersion.

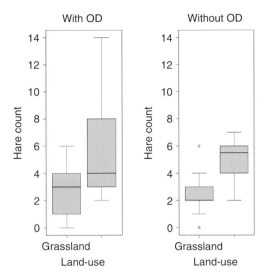

FIGURE 14.1   Hare counts by land-use with and without overdispersion (OD). Overdispersion was caused by site-specific differences in hare density.

```
n.site <- 10
x <- gl(n = 2, k = n.site, labels = c("grassland", "arable"))
eps <- rnorm(2*n.site, mean = 0, sd = 0.5)                    # Normal random effect
lambda.OD <- exp(0.69 +(0.92*(as.numeric(x)-1) + eps) )
lambda.Poisson <- exp(0.69 +(0.92*(as.numeric(x)-1)) )# For comparison
```

We add the noise that comes from a Poisson and inspect the hare counts we've generated (Fig. 14.1):

```
C.OD <- rpois(n = 2*n.site, lambda = lambda.OD)
C.Poisson <- rpois(n = 2*n.site, lambda = lambda.Poisson)

par(mfrow = c(1,2))
boxplot(C.OD ~ x, col = "grey", xlab = "Land-use", main = "With OD", ylab = "Hare
count", las = 1, ylim = c(0, max(C.OD)))
boxplot(C.Poisson ~ x, col = "grey", xlab = "Land-use", main = "Without OD", ylab =
"Hare count", las = 1, ylim = c(0, max(C.OD)) )
```

14.1.3 Analysis Using R

We conduct a classical analysis of the overdispersed data once without and then with correction for overdispersion (using `glm(, family = quasi...)`).

```
glm.fit.no.OD <- glm(C.OD ~ x, family = poisson)
glm.fit.with.OD <- glm(C.OD ~ x, family = quasipoisson)
summary(glm.fit.no.OD)
```

```
summary(glm.fit.with.OD)
anova(glm.fit.no.OD, test = "Chisq")
anova(glm.fit.with.OD, test = "F")

> summary(glm.fit.no.OD)

Call:
glm(formula = C.OD ~ x, family = poisson)

[ ... ]

Coefficients:
             Estimate   Std. Error   z value   Pr(>|z|)
(Intercept)    0.9933       0.1925     5.161   2.46e-07 ***
xarable        0.7646       0.2330     3.282   0.00103  **
- - -
[ ... ]

(Dispersion parameter for poisson family taken to be 1)

[ ... ]

> summary(glm.fit.with.OD)

Call:
glm(formula = C.OD ~ x, family = quasipoisson)

[ ... ]

Coefficients:
             Estimate   Std. Error   t value    Pr(>|t|)
(Intercept)    0.9933       0.2596     3.826     0.00124  **
xarable        0.7646       0.3143     2.433     0.02563  *
- - -
[ ... ]

(Dispersion parameter for quasipoisson family taken to be 1.819569)

    Null deviance:  44.198  on 19   degrees of freedom
Residual deviance:  32.627  on 18   degrees of freedom
[ ... ]

> anova(glm.fit.no.OD, test = "Chisq")
Analysis of Deviance Table

Model: poisson, link: log

[ ... ]

       Df   Deviance   Resid. Df   Resid. Dev   P(>|Chi|)
NULL                        19         44.198
x       1    11.571          18         32.627       0.001
> anova(glm.fit.with.OD, test = "F")
Analysis of Deviance Table
```

```
Model: quasipoisson, link: log

[ ... ]

      Df  Deviance  Resid. Df  Resid. Dev      F    Pr(>F)
NULL                    19       44.198
x      1    11.571       18       32.627  6.3591  0.02132 *
```

Thus, the parameter estimates don't change when accounting for overdispersion, but tests and standard errors do.

14.1.4 Analysis Using WinBUGS

In WinBUGS, it is easy to get from the simple Poisson t-test with homogeneous (Poisson) variance in the last chapter to the overdispersed Poisson t-test represented by the Poisson-lognormal model.

```
# Define model
sink("Poisson.OD.t.test.txt")
cat("
model {

# Priors
 alpha ~ dnorm(0,0.001)
 beta ~ dnorm(0,0.001)
 sigma ~ dunif(0, 10)
 tau <- 1 / (sigma * sigma)

# Likelihood
 for (i in 1:n) {
    C.OD[i] ~ dpois(lambda[i])
    log(lambda[i]) <- alpha + beta *x[i] + eps[i]
    eps[i] ~ dnorm(0, tau)
 }
}
",fill=TRUE)
sink()

# Bundle data
win.data <- list(C.OD = C.OD, x = as.numeric(x)-1, n = length(x))

# Inits function
inits <- function(){ list(alpha=rlnorm(1), beta=rlnorm(1), sigma = rlnorm(1))}

# Parameters to estimate
params <- c("lambda","alpha", "beta", "sigma")
```

Note that as soon as we start estimating variances (here, of the overdispersion effects eps), we need longer chains.

```
# MCMC settings
nc <- 3             # Number of chains
ni <- 3000          # Number of draws from posterior per chain
nb <- 1000          # Number of draws to discard as burn-in
nt <- 5             # Thinning rate

# Start Gibbs sampling
out <- bugs(data=win.data, inits=inits, parameters.to.save=params,
model.file="Poisson.OD.t.test.txt", n.thin=nt, n.chains=nc, n.burnin=nb, n.iter=ni,
debug = TRUE)

print(out, dig = 3)
```

14.2 ZERO-INFLATION

14.2.1 Introduction

Zero-inflation can be called a specific form of overdispersion and is frequently found in count data. It means that there are more zeroes than expected under the assumed (e.g., Poisson or binomial) distribution. In the context of our hare counts, a typical explanation for excess zeroes is that some sites are simply not suitable for hares, such as paved parking lots, roof tops, or lakes; hence, resulting counts must be zeroes. In the remaining suitable sites, counts vary according to the assumed distribution. Thus, we may imagine a sequential genesis of zero-inflated counts: first, Nature determines whether a site may be occupied at all, and second, she selects the counts for those that are habitable in principle. Regression models that account for this kind of overdispersion are often called zero-inflated Poisson (ZIP) or zero-inflated binomial (ZIB) models.

A ZIP model for count C_i at site i can be written algebraically like this:

$$w_i \sim \text{Bernoulli}(\psi_i) \qquad \text{Suitability of a site} \qquad (14.1)$$

$$C_i \sim \text{Poisson}(w_i \times \lambda_i) \qquad \text{Observed counts} \qquad (14.2)$$

For each site i, Nature flips a coin that lands heads (i.e., $w_i = 1$) with probability ψ_i. We can't observe w_i perfectly, i.e., it is a latent or random effect. Only for sites with $w_i = 1$, Nature then rolls her Poisson (λ_i) die to determine the count C_i at that site. For sites with $w_i = 0$, the Poisson mean is $0 \times \lambda_i = 0$ and the corresponding Poisson die produces zero counts only.

We see that a ZIP model simply represents a set of two coupled GLMs: the logistic regression describes the suitability in principle of a site while the Poisson regression describes the variation of counts among suitable sites, i.e., those with $w_i = 1$. All the usual GLM features apply, and in particular, both the Bernoulli and the Poisson parameter can be expressed as a function of covariates on the link scale. These covariates may or may not be the same for both regressions.

I make four notes on the ZIP model in our context of hare counts: First, the model allows for two entirely different kinds of zero counts; those coming from the Bernoulli and those from the Poisson process. The former are zero counts at unsuitable sites while the latter are due to Poisson chance, i.e., for them, Nature's Poisson die happened to yield a zero. The actual distribution of an organism, i.e., the proportion of sites that is occupied (has nonzero counts), is a function of both processes. Hence, it would be wrong to say that Eqn. 14.1 describes the distribution and Eqn. 14.2 the abundance.

Second, the above ZIP model is a hierarchical, or random-effects, model with binary instead of normal random effects. It is an example of the kind of nonstandard GLMMs that are featured extensively in Chapters 20 and 21. There, we will see the site-occupancy species distribution model, another kind of zero-inflated GLM, but one where a Bernoulli or binomial distribution is zero-inflated with another Bernoulli, so we get a zero-inflated binomial (ZIB) model.

Third, some authors advocate ZIP models widely for inference about count data (Martin et al., 2005; Joseph et al., 2009). However, on ecological grounds, they appear most adequate in situations where *unknown* environmental covariates determine the suitability of a site. If covariates are known and have been measured, they are probably best added to the linear model for the Poisson mean. Distribution, or occurrence, is fundamentally a function of abundance, i.e., a species occurs at all sites where abundance is greater than zero. It appears contrived to model distribution completely separately from abundance.

Fourth, there is a variant of a ZIP model called the hurdle model (Zeileis et al., 2008), where the first step in the hierarchical genesis of the counts is assumed to be the same as in a ZIP model, i.e., $w_i \sim$ Bernoulli(ψ_i). But then, counts at suitable sites (i.e., with $w_i = 1$) are modeled as coming from a zero-truncated Poisson distribution, i.e., a Poisson for values excluding zero. Hurdles (thresholds) other than zero are also possible. Superficially, this model may appear "better" than a ZIP model because it only allows one kind of zero: that coming from the Bernoulli process. However, it posits that all sites that are suitable in principle *will* be occupied and have a count greater than 0. This is not sensible biologically because, in reality, a suitable site may well be unoccupied as a result of local extinction, dispersal limitation, or some other reason.

14.2.2 Data Generation

We generate the simplest kind of zero-inflated count data for our (Poisson) hare example. We assume different densities in arable and grassland areas and a constant zero inflation, i.e., a single value of ψ for all sites, regardless of land-use or other environmental covariates.

```
psi <- 0.8
n.site <- 20
x <- gl(n = 2, k = n.site, labels = c("grassland", "arable"))
```

For each site, we flip a coin to determine its suitability and store the result in the latent state w_i.

```
w <- rbinom(n = 2*n.site, size = 1, prob = psi)
```

We assume identical effects of arable and grass as before and generate expected counts at suitable sites as before.

```
lambda <- exp(0.69 +(0.92*(as.numeric(x)-1)) )
```

We then add up (actually, multiply) the effects of both processes (Bernoulli and Poisson) and inspect the counts we've generated. Note how all counts at unsuitable sites (with $w_i = 0$) are zero.

```
C <- rpois(n = 2*n.site, lambda = w *lambda)
cbind(x, w, C)
```

14.2.3 Analysis Using R

A wide range of ZIP and related models can be fitted in R using the function `zeroinfl()` in package pscl; see, for instance, Zeileis et al. (2008). We load that package and fit the simplest possible ZIP model.

```
library(pscl)
fm <- zeroinfl(C ~ x | 1, dist = "poisson")

summary(fm)
> summary(fm)

Call:
zeroinfl(formula = C ~ x | 1, dist = "poisson")

Count model coefficients (poisson with log link):
            Estimate  Std. Error  z value   Pr(>|z|)
(Intercept)   0.7441      0.1773    4.197   2.71e-05 ***
xarable       0.8820      0.2095    4.209   2.56e-05 ***

Zero-inflation model coefficients (binomial with logit link):
            Estimate  Std. Error  z value   Pr(>|z|)
(Intercept)  -1.6786      0.4943   -3.396   0.000684 ***
- - -
Signif. codes: 0'***' 0.001 '**' 0.01 '*' 0.05 '.' 0.1 ' ' 1

Number of iterations in BFGS optimization: 8
Log-likelihood: -78.4 on 3 Df
```

Because of sampling and estimation error, the coefficients for the count model (corresponding to Eqn. 14.2) may not always be very close to the input values. Also note that what pscl calls the coefficient in the zero-inflation model corresponds to $1 - \psi$ in Eqn. 14.1. Typing plogis (-1.6786) in R convinces us that the function is doing what it should do.

14.2.4 Analysis Using WinBUGS

Next, the solution in WinBUGS. As always, the elementary manner of model specification using the BUGS language makes it very clear what model is fitted. To make the parameter estimates directly comparable, we also add a line that computes the logit of the zero-inflation parameter in R from the parameter ψ that we use here.

```
# Define model
sink("ZIP.txt")
cat("
model {

# Priors
 psi ~ dunif(0,1)
 alpha ~ dnorm(0,0.001)
 beta ~ dnorm(0,0.001)

# Likelihood
 for (i in 1:n) {
    w[i] ~ dbern(psi)
    C[i] ~ dpois(eff.lambda[i])
    eff.lambda[i] <- w[i]*lambda[i]
    log(lambda[i]) <- alpha + beta *x[i]
 }

# Derived quantity
 R.lpsi <- logit(1-psi)
}
",fill=TRUE)
sink()

# Bundle data
win.data <- list(C = C, x = as.numeric(x)-1, n = length(x))

# Inits function
inits <- function(){ list(alpha=rlnorm(1), beta=rlnorm(1), w = rep(1, 2*n.site))}
```

We will also estimate the latent state w_i, i.e., the intrinsic suitability for brown hares at each site.

```
# Parameters to estimate
params <- c("lambda","alpha", "beta", "w", "psi", "R.lpsi")

# MCMC settings (need fairly long chains)
nc <- 3              # Number of chains
ni <- 50000          # Number of draws from posterior per chain
nb <- 10000          # Number of draws to discard as burn-in
nt <- 4              # Thinning rate

# Start WinBUGS
out <- bugs(data=win.data, inits=inits, parameters.to.save=params,
model.file="ZIP.txt", n.thin=nt,n.chains=nc,n.burnin=nb, n.iter=ni, debug = TRUE)

print(out, dig = 3)
> print(out, dig = 3)
Inference for Bugs model at "ZIP.txt", fit using WinBUGS,
 3 chains, each with 50000 iterations (first 10000 discarded), n.thin = 4
 n.sims = 30000 iterations saved
```

	mean	sd	2.5%	25%	50%	75%	97.5%	Rhat	n.eff
[...]									
alpha	0.733	0.175	0.384	0.617	0.736	0.853	1.066	1.001	11000
beta	0.884	0.208	0.479	0.744	0.883	1.024	1.300	1.001	30000
[...]									
psi	0.827	0.064	0.689	0.786	0.831	0.873	0.936	1.001	30000
R.lpsi	-1.635	0.484	-2.684	-1.931	-1.596	-1.302	-0.795	1.001	30000

We find pretty similar estimates between R and WinBUGS.

14.3 OFFSETS

14.3.1 Introduction

In a Poisson GLM, we assume that the expected counts are adequately described by the effect of the covariates in the model. However, frequently, we have that the "counting window" is not constant, e.g., that study areas don't have the same size or, in temporal samples, that the duration of counting periods differ. To account for this known component of variation in the conditional Poisson mean, we define the log of the size of the "counting window" (study area size, count duration) as an *offset*. Effectively, we then model density as a response.

Let's consider this for the hare counts using algebra. The Poisson GLM is $C_i \sim \text{Poisson}(\lambda_i)$, i.e., hare counts C_i are conditionally distributed as Poisson with expected count λ_i. When study areas differ in size, we have $C_i \sim \text{Poisson}(A_i * \lambda_i)$, where A_i is the area of study area i. Therefore, the linear predictor becomes $\log(A_i * \lambda_i) = \log(A_i) + \log(\lambda_i)$. If we also wish to model a covariate x into the mean, we get $\log(A_i * \lambda_i) = \log(A_i) + \alpha + \beta * x_i$. This is equivalent to forcing the coefficient of log(area) to be equal to 1. That is, we effectively fit the model $\log(A_i * \lambda_i) = \beta_0 * \log(A_i) + \alpha + \beta * x_i$ with $\beta_0 = 1$. The offset compensates for the additional and known variation in the response resulting from differing study area size.

14.3.2 Data Generation

```
n.site <- 10
A <- runif(n = 2*n.site, 2,5)    # Areas range in size from 2 to 5 km2
x <- gl(n = 2, k = n.site, labels = c("grassland", "arable"))
linear.predictor <- log(A) + 0.69 +(0.92*(as.numeric(x)-1))
lambda <- exp(linear.predictor)
C <- rpois(n = 2*n.site, lambda = lambda)  # Add Poisson noise
```

14.3.3 Analysis Using R

We use R for an analysis with and without consideration of the differing areas.

```
glm.fit.no.offset <- glm(C ~ x, family = poisson)
glm.fit.with.offset <- glm(C ~ x, family = poisson, offset = log(A))
summary(glm.fit.no.offset)
summary(glm.fit.with.offset)
anova(glm.fit.with.offset, test = "Chisq")            # LRT
```

Comparing the residual deviance of the two models makes clear that specification of an offset represents a sort of correction for a systematic kind of overdispersion.

14.3.4 Analysis Using WinBUGS

Note how simple it is in WinBUGS to jump from one kind of analysis for the hare counts to another.

```
# Define model
sink("Offset.txt")
cat("
model {

# Priors
 alpha ~ dnorm(0,0.001)
 beta ~ dnorm(0,0.001)

# Likelihood
 for (i in 1:n) {
```

```
    C[i] ~ dpois(lambda[i])
    log(lambda[i]) <- 1 * logA[i] + alpha + beta *x[i]    # Note offset
  }
}
",fill=TRUE)
sink()

# Bundle data
win.data <- list(C = C, x = as.numeric(x)-1, logA = log(A), n = length(x))

# Inits function
inits <- function(){ list(alpha=rlnorm(1), beta=rlnorm(1))}

# Parameters to estimate
params <- c("lambda","alpha", "beta")

# MCMC settings
nc <- 3              # Number of chains
ni <- 1100           # Number of draws from posterior
nb <- 100            # Number of draws to discard as burn-in
nt <- 2              # Thinning rate

# Start Gibbs sampling
out <- bugs(data=win.data, inits=inits, parameters.to.save=params,
model.file="Offset.txt", n.thin=nt,n.chains=nc,n.burnin=nb, n.iter=ni, debug =
TRUE)

print(out, dig = 3)
```

14.4 SUMMARY

Overdispersion, zero-inflation, and offsets are important GLM topics. The specification of the associated models in WinBUGS is fairly easy and clarifies the actual meaning of these three topics. This is not usually the case when fitting these models in a canned routine in R or another software. Thus, this is another example of where the simple model specification in the BUGS language enforces an understanding of the fitted model that is easily lost in other stats packages.

EXERCISES

1. *Estimating a coefficient for an offset covariate*: Forcing the coefficient of area to be 1 implies the assumption that hare density is unaffected by area. However, we can also estimate a coefficient for log(area) rather than setting it to 1. For instance, in real life, density may well differ between

large and small areas, perhaps because predator density is also related to area. This hypothesis could be tested by fitting a coefficient for log(area) and seeing whether it differs from 1. Adapt the WinBUGS code to achieve this. A Bayesian analysis is extremely suited for this type of question (i.e., to test whether a parameter has a value other than what it is expected to be under a certain hypothesis).

2. *Swiss hare data*: Fit a model that contains both overdispersion, modeled as log-normal random effect, and an offset of log(area), when estimating the difference in mean density between arable and grassland study areas in one year, e.g., 2000 (use a single count per year and area). In a variant of that analysis, estimate the coefficient of log(area) to test whether hare density depends on area.

Poisson ANCOVA

15.1 INTRODUCTION

We may call a Poisson analysis of covariance (ANCOVA) a Poisson regression with both discrete and continuous covariates. In most practical applications, Poisson models will have several covariates and of both types. Therefore, here we look at this important variety of a generalized linear model (GLM). To stress the similarity with the normal linear case, we only slightly alter the inferential setting sketched in Chapter 11. We assume that instead of measuring body mass in asp vipers in three populations in the Pyrenees, Massif Central, and the Jura mountains, leading to a normal model, we had instead assessed ectoparasite load in a dragonfly, the Sombre Goldenring (Fig. 15.1), leading to a Poisson model. We are particularly interested in whether there are more or less little red mites on dragonflies of different size (expressed as wing length) and whether this relationship differs among the three mountain ranges. (Actually, dragonflies don't vary that much in body size, but let's assume there is sufficient variation to make such a study worthwhile.)

FIGURE 15.1   Sombre Goldenring (*Cordulegaster bidentata*), Switzerland, 1995. (*Photo F. Labhardt*)

We will fit the following model to mite count C_i on individual i:

1. Distribution: $C_i \sim \text{Poisson}(\lambda_i)$
2. Link function: log, i.e., $\log(\lambda_i) = \log(E(C_i)) = \text{linear predictor}$
3. Lin. predictor: $\log(\lambda_i) = \alpha_{\text{Pyr}} + \beta_1 * x_{\text{MC}} + \beta_2 * x_{\text{Jura}} + \beta_3 * x_{\text{wing}} + \beta_4 * x_{\text{wing}} * x_{\text{MC}} + \beta_5 * x_{\text{wing}} * x_{\text{Jura}}$

Note the great similarity between this model and the one we fitted to the mass of asps in Chapter 11. Apart from the link function, the main other difference is simply that for this model we don't have a dispersion term; the Poisson already comes with a built-in variability. We could model overdispersion in mite counts by using a Poisson-lognormal formulation as in the previous chapter, but we omit such added complexity here.

15.2 DATA GENERATION

We assemble the data set.

```
n.groups <- 3
n.sample <- 100
n <- n.groups * n.sample
```

```
x <- rep(1:n.groups, rep(n.sample, n.groups))    # Population indicator
pop <- factor(x, labels = c("Pyrenees", "Massif Central", "Jura"))

length <- runif(n, 4.5, 7.0)                      # Wing length (cm)
length <- length - mean(length)                   # Centre by subtracting the mean
```

We build the design matrix of an interactive combination of length and population:

```
Xmat <- model.matrix(~ pop*length)
print(Xmat, dig = 2)
```

Select the parameter values, i.e., choose values for α_{Pyr}, β_1, β_2, β_3, β_4, β_5

```
beta.vec <- c(-2, 1, 2, 5, -2, -7)
```

Here's the recipe for assembling the mite counts in three steps:

1. we add up all components of the linear model to get the linear predictor, which is the expected mite count on a (natural) log scale,
2. we exponentiate to get the actual value of the expected mite count, and
3. we add Poisson noise.

We again obtain the value of the linear predictor by matrix multiplication of the design matrix (Xmat) and the parameter vector (beta.vec) (As in the viper example, don't be too harsh on me in case we achieve unnatural levels of parasitation ...):

```
lin.pred <- Xmat[,] %*% beta.vec      # Value of lin.predictor
lambda <- exp(lin.pred)               # Poisson mean
C <- rpois(n = n, lambda = lambda)    # Add Poisson noise

# Inspect what we've created
par(mfrow = c(2,1))
hist(C, col = "grey", breaks = 30, xlab = "Parasite load", main = "", las = 1)
plot(length, C, pch = rep(c("P","M","J"), each=n.sample), las = 1, col =
rep(c("Red","Green","Blue"), each=n.sample), ylab = "Parasite load", xlab = "Wing
length", cex = 1.2)
par(mfrow = c(1,1))
```

We have created a data set where parasite load increases with wing length in the South (Pyrenees, Massif Central) but decreases in the North (Jura mountains); see Fig. 15.2.

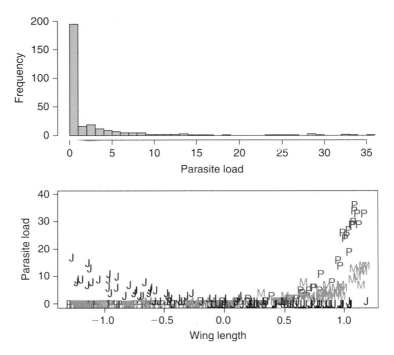

FIGURE 15.2   Top: Frequency distribution of ectoparasite load in Sombre Goldenrings. Bottom: Relationship between parasite load and wing length (deviation from mean, in cm) in three mountain ranges (P – Pyrenees, M – Massif Central, J – Jura mountains).

15.3 ANALYSIS USING R

Again, the R code to fit the model is very parsimonious and the estimates resemble the input values reasonably well (coefficients can be compared directly with beta.vec). Obviously, with larger sample sizes the correspondence would be better still (you could try that out, e.g., by setting n.sample = 1000).

```
summary(glm(C ~ pop * length, family = poisson))
beta.vec

> summary(glm(C ~ pop * length, family = poisson))

Call:
glm(formula = C ~ pop * length, family = poisson)

Deviance Residuals:
    Min       1Q   Median       3Q      Max
-2.1471  -0.6680  -0.1961   0.2141   2.8843
```

```
Coefficients:
                         Estimate   Std. Error   z value    Pr(>|z|)
(Intercept)              -1.8540       0.2554     -7.260    3.87e-13 ***
popMassif Central         0.6386       0.3463      1.844      0.0652 .
popJura                   1.8046       0.2851      6.330    2.46e-10 ***
length                    4.8199       0.2505     19.244     < 2e-16 ***
popMassif Central:length -1.6938       0.3516     -4.817    1.46e-06 ***
popJura:length           -6.9059       0.2859    -24.156     < 2e-16 ***
---
Signif. codes:  0 '***' 0.001 '**' 0.01 '*' 0.05 '.' 0.1 ' ' 1

(Dispersion parameter for poisson family taken to be 1)

    Null deviance: 2402.99  on 299  degrees of freedom
Residual deviance:  205.35  on 294  degrees of freedom
AIC: 684.25

Number of Fisher Scoring iterations: 5

> beta.vec
[1] -2 1 2 5 -2 -7
```

Don't forget that the difference between the coefficients and the beta vector is due to the combined effect of sampling and estimation error. The former means that due to natural variation, 300 dragonflies sampled from large populations can not possibly represent the population values perfectly.

15.4 ANALYSIS USING WinBUGS

15.4.1 Fitting the Model

We simply adapt the code for the normal linear case (Chapter 11) to the Poisson case and again fit a reparameterized model with three separate log-linear regressions.

```
# Define model
sink("glm.txt")
cat("
model {

# Priors
  for (i in 1:n.groups){
    alpha[i] ~ dnorm(0, 0.01)        # Intercepts
    beta[i] ~ dnorm(0, 0.01)         # Slopes
  }
```

```
# Likelihood
  for (i in 1:n) {
      C[i] ~ dpois(lambda[i])          # The random variable
      lambda[i] <- exp(alpha[pop[i]] + beta[pop[i]]* length[i])
  }    # Note double-indexing: alpha[pop[i]]

# Derived quantities
# Recover effects relative to baseline level (no. 1)
  a.effe2 <- alpha[2] - alpha[1]       # Intercept Massif Central vs. Pyr.
  a.effe3 <- alpha[3] - alpha[1]       # Intercept Jura vs. Pyr.
  b.effe2 <- beta[2] - beta[1]         # Slope Massif Central vs. Pyr.
  b.effe3 <- beta[3] - beta[1]         # Slope Jura vs. Pyr.

# Custom test
  test1 <- beta[3] - beta[2]                 # Slope Jura vs. Massif Central
  }
",fill=TRUE)
sink()

# Bundle data
win.data <- list(C = C, pop = as.numeric(pop), n.groups = n.groups, length =
length, n = n)

# Inits function
inits <- function(){list(alpha=rlnorm(n.groups,3,1), beta=rlnorm(n.groups,2,1))}

# Parameters to estimate
params <- c("alpha", "beta", "a.effe2", "a.effe3", "b.effe2", "b.effe3", "test1")

# MCMC settings
ni <- 4500
nb <- 1500
nt <- 5
nc <- 3

# Start Gibbs sampling
out <- bugs(data = win.data, inits = inits, parameters.to.save = params, model.file
= "glm.txt", n.thin = nt, n.chains = nc, n.burnin = nb, n.iter = ni, debug = TRUE)
```

This is still a simple model that converges fairly rapidly. We inspect the results and compare them with "truth" in the data-generating random process, as well as with the inference from R's function `glm()`.

```
print(out, dig = 3)         # Bayesian analysis
beta.vec                    # Truth in data-generating process
print(glm(C ~ pop * length, family = poisson)$coef, dig = 4) # The ML solution
```

Remember that alpha[1] and beta[1] in WinBUGS correspond to the intercept and the length main effect in the analysis in R and a.effe2, a.effe3. b.effe2, b.effe3 to the remaining terms of the analysis in R. To ease comparison, these terms are shown in boldface.

```
> print(out, dig = 3)                      # Bayesian analysis
Inference for Bugs model at "glm.txt", fit using WinBUGS.
 3 chains, each with 4500 iterations (first 1500 discarded), n.thin = 5
 n.sims = 1800 iterations saved
               mean      sd    2.5%     25%     50%     75%   97.5%    Rhat   n.eff
alpha[1]     -1.844   0.261  -2.384  -2.016  -1.852  -1.666  -1.360   1.017    120
alpha[2]     -1.252   0.233  -1.744  -1.399  -1.235  -1.089  -0.821   1.016    130
alpha[3]     -0.050   0.130  -0.321  -0.136  -0.046   0.037   0.186   1.002   1400
beta[1]       4.808   0.258   4.329   4.628   4.809   4.974   5.329   1.015    140
beta[2]       3.159   0.245   2.712   2.987   3.141   3.322   3.661   1.015    140
beta[3]      -2.087   0.141  -2.367  -2.176  -2.083  -1.987  -1.811   1.002    940
a.effe2       0.592   0.344  -0.063   0.356   0.590   0.825   1.264   1.005    440
a.effe3       1.793   0.291   1.231   1.597   1.793   1.989   2.371   1.015    140
b.effe2      -1.649   0.352  -2.324  -1.887  -1.647  -1.403  -0.970   1.004    530
b.effe3      -6.895   0.296  -7.470  -7.092  -6.891  -6.690  -6.336   1.011    190
test1        -5.246   0.282  -5.823  -5.439  -5.240  -5.044  -4.720   1.015    140
deviance    678.382   3.450 673.597 675.800 677.800 680.200 686.202   1.003   1200

For each parameter, n.eff is a crude measure of effective sample size,
and Rhat is the potential scale reduction factor (at convergence, Rhat=1).

DIC info (using the rule, pD = Dbar-Dhat)
pD = 6.1 and DIC = 684.5
DIC is an estimate of expected predictive error (lower deviance is better).

> beta.vec# Truth in data-generating process
[1] -2  1  2  5 -2 -7

> print(glm(C ~ pop * length, family = poisson)$coef, dig = 4) # The ML solution
              (Intercept)            popMassif Central                    popJura
                  -1.8540                        0.6386                     1.8046
              length popMassif Central:length         popJura:length
                  4.8199                       -1.6938                    -6.9059
```

As expected, we find fairly concurrent estimates between the two modes of inference.

15.4.2 Forming Predictions

Finally, let's summarize our main findings from the analysis in a graph. We illustrate the posterior distribution of the relationship between mite load and wing length for each of the three study areas. To do that,

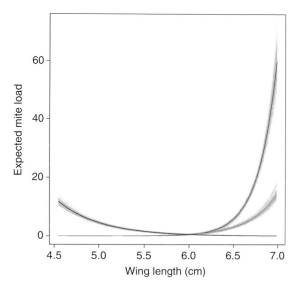

FIGURE 15.3 Predicted relationship between mite load and the size of a dragonfly in three mountain ranges: Pyrenees (red), Massif Central (green), and Jura mountains (blue). Shaded grey lines illustrate the uncertainty in these relationships based on a random sample of 100 from the posterior distribution and colored lines represent posterior means.

we predict mite load for 100 dragonflies in each of the three mountain ranges and plot these estimates along with their uncertainty. We compute the predicted relationship between mite count and wing-length for a sample of 100 of all Markov chain Monte Carlo (MCMC) draws of the involved parameters and plot that (Fig. 15.3).

```
# Create a vector with 100 wing lengths
original.wlength <- sort(runif(100, 4.5, 7.0))
wlength <- original.wlength - 5.75     # 5.75 is approximate mean (correct would be
that from the original data really)

# Create matrices to contain prediction for each winglength and MCMC iteration
sel.sample <- sample(1:1800, size = 100)
mite.load.Pyr <- mite.load.MC <- mite.load.Ju <- array(dim = c(100, 100))

# Fill in these vectors: this is clumsy, but it works
for(i in 1:100) {
    for(j in 1:100)   {
      mite.load.Pyr[i,j] <- exp(out$sims.list$alpha[sel.sample[j],1] +
out$sims.list$beta[sel.sample[j],1] * wlength[i])
      mite.load.MC[i,j] <- exp(out$sims.list$alpha[sel.sample[j],2] +
out$sims.list$beta[sel.sample[j],2] * wlength[i])
      mite.load.Ju[i,j] <- exp(out$sims.list$alpha[sel.sample[j],3] +
out$sims.list$beta[sel.sample[j],3] * wlength[i])
```

```
    }
  }

matplot(original.wlength, mite.load.Pyr, col = "grey", type = "l", las = 1, ylab =
"Expected mite load", xlab = "Wing length (cm)")
for(j in 1:100){
    points(original.wlength, mite.load.MC[,j], col = "grey", type = "l")
    points(original.wlength, mite.load.Ju[,j], col = "grey", type = "l")
  }
points(original.wlength, exp(out$mean$alpha[1] + out$mean$beta[1] * wlength), col =
"red", type = "l", lwd = 2)
points(original.wlength, exp(out$mean$alpha[2] + out$mean$beta[2] * wlength), col =
"green", type = "l", lwd = 2)
points(original.wlength, exp(out$mean$alpha[3] + out$mean$beta[3] * wlength), col =
"blue", type = "l", lwd = 2)
```

I find this a rather nice plot, but if an editor asks for conventional 95% credible intervals instead, you can give the results of this code:

```
LCB.Pyr <- apply(mite.load.Pyr, 1, quantile, prob=0.025)
UCB.Pyr <- apply(mite.load.Pyr, 1, quantile, prob=0.975)
LCB.MC <- apply(mite.load.MC, 1, quantile, prob=0.025)
UCB.MC <- apply(mite.load.MC, 1, quantile, prob=0.975)
LCB.Ju <- apply(mite.load.Ju, 1, quantile, prob=0.025)
UCB.Ju <- apply(mite.load.Ju, 1, quantile, prob=0.975)

mean.rel <- cbind(exp(out$mean$alpha[1] + out$mean$beta[1] * wlength),
exp(out$mean$alpha[2] + out$mean$beta[2] * wlength), exp(out$mean$alpha[3] +
out$mean$beta[3] * wlength))
covar <- cbind(original.wlength, original.wlength, original.wlength)

matplot(original.wlength, mean.rel, col = c("red", "green", "blue"), type = "l",
lty = 1, lwd = 2, las = 1, ylab = "Expected mite load", xlab = "Wing length (cm)")
points(original.wlength, LCB.Pyr, col = "grey", type = "l", lwd = 1)
points(original.wlength, UCB.Pyr, col = "grey", type = "l", lwd = 1)
points(original.wlength, LCB.MC, col = "grey", type = "l", lwd = 1)
points(original.wlength, UCB.MC, col = "grey", type = "l", lwd = 1)
points(original.wlength, LCB.Ju, col = "grey", type = "l", lwd = 1)
points(original.wlength, UCB.Ju, col = "grey", type = "l", lwd = 1)
```

15.5 SUMMARY

We have generalized the general linear model from the normal to the Poisson case to model the effects on grouped counts of a continuous covariate. The changes involved in doing so in WinBUGS were minor, and the inclusion of further covariates is straightforward. The Poisson ANCOVA is an important intermediate step for your understanding of the Poisson generalized linear mixed model.

EXERCISES

1. *Multiple Poisson regression*: Invent a new covariate called $x_{nonsense}$, for instance, and fit this model: $y_i = \alpha_j + \beta_j * x_{wing} + \delta * x_{nonsense}$. You can simply create values for $x_{nonsense}$ by sampling a normal or uniform random variable; you don't need to assemble a new data set.

2. *Polynomial Poisson regression*: In addition to the linear effect (on a log-scale) of wing length, check for a quadratic effect also. You may or may not reassemble the data to include that effect.

3. *Swiss hare data*: Take a single count per year and site and fit a Poisson ANCOVA to the counts by expressing counts as a function of site and year. You may ignore the variable site area or incorporate this source of variation by fitting an offset.

Poisson Mixed-Effects Model (Poisson GLMM)

16.1 INTRODUCTION

The Poisson generalized linear mixed model (GLMM) is an extension of the Poisson generalized linear model (GLM) to include at least one additional source of random variation over and above the random variation intrinsic to a Poisson distribution. Here, we adopt a Poisson GLMM to analyze a set of long-term population surveys of Red-backed shrikes (Fig. 16.1).

We assume that pair counts over 30 years were available in each of 16 shrike populations (again, the balanced design is for convenience only). Our intent is to model population trends. First, we write down the random-coefficients model without correlation between the intercepts and slopes. This model is very similar to that for the normal linear case that we examined extensively in Chapter 12. Thanks to how we specify models in the BUGS language, this similarity is more evident than when fitting the model with a canned routine such as in R.

FIGURE 16.1   Male Red-backed shrike (*Lanius collurio*), Switzerland, 2004. (*Photo A. Saunier*)

We model C_i, the number pairs of Red-backed shrikes counted in year i in study area j:

1. Distribution: $C_i \sim \text{Poisson}(\lambda_i)$
2. Link function: log, i.e., $\log(\lambda_i) = \log(E(C_i)) = $ linear predictor
3. Linear predictor: $\alpha_{j(i)} + \beta_{j(i)} * x_i$
4. Submodel for parameters/distribution of random effects:

$$\alpha_j \sim \text{Normal}\,(\mu_\alpha, \sigma_\alpha^2)$$

$$\beta_j \sim \text{Normal}\,(\mu_\beta, \sigma_\beta^2)$$

So the GLMM is just the same as a simple GLM, but with the added submodels for the log-linear intercept and slope parameters that we use to describe the population trends. We don't add a year-specific "residual" to the linear predictor. This could be done to account for random year effects and would be a first step towards modeling serial autocorrelation, e.g., by imposing an autoregressive structure on successive random year effects (Littell et al., 2008), but we omit this added complexity here. Also, we don't model a correlation between intercepts and slopes.

In conducting this analysis, we implicitly make one of two assumptions. Either we assume that we find all shrike pairs in every year and study area or we assume that at least the proportion of pairs overlooked does not vary among years or study areas in a systematic way, i.e., that there is no time trend in detection probability. Otherwise, the interpretation of these data would be difficult or impossible. For an alternative protocol for data collection and associated analysis method that doesn't require these potentially restrictive assumptions, see the binomial mixture model in Chapter 21.

16.2 DATA GENERATION

We generate data under the random-coefficients model without correlation between the intercepts and slopes.

```
n.groups <- 16
n.years <- 30
n <- n.groups * n.years
pop <- gl(n = n.groups, k = n.years)
```

We standardize the year covariate to a range from zero to one.

```
original.year <- rep(1:n.years, n.groups)
year <-  (original.year-1)/29
```

We build a design matrix without the intercept.

```
Xmat <- model.matrix(~pop*year-1-year)
print(Xmat[1:91,], dig = 2)              # Print top 91 rows
```

Next, we draw the intercept and slope parameter values from their normal distributions, whose hyperparameters we need to select.

```
intercept.mean <- 3       # Choose values for the hyperparams
intercept.sd <- 1
slope.mean <- -2
slope.sd <- 0.6
intercept.effects<-rnorm(n=n.groups, mean=intercept.mean, sd=intercept.sd)
slope.effects <- rnorm(n = n.groups, mean = slope.mean, sd = slope.sd)
all.effects <- c(intercept.effects, slope.effects)   # Put them all together
```

We assemble the counts C_i by first computing the linear predictor, then exponentiating it and finally adding Poisson noise. Then, we look at the data.

```
lin.pred <- Xmat[,] %*% all.effects       # Value of lin.predictor
C <- rpois(n = n, lambda = exp(lin.pred))  # Exponentiate and add Poisson noise
hist(C, col = "grey")                      # Inspect what we've created
```

We use a lattice graph to plot the shrike counts against time for each population (Fig. 16.2).

```
library("lattice")
xyplot(C ~ original.year | pop, ylab = "Red-backed shrike counts", xlab = "Year")
```

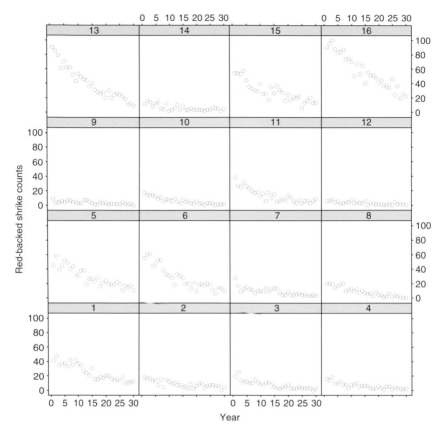

FIGURE 16.2   Trellis plot of pair counts in 16 populations of Red-backed shrikes over 30 years.

16.3 ANALYSIS UNDER A RANDOM-COEFFICIENTS MODEL

We could analyze the shrike counts under the assumption that all shrike populations have the same trend, but at different levels, corresponding to a random-intercepts model, as we did for the normal case (see Section 12.3). However, we only show here the analysis under the random-coefficients model. We assume that each population has a specific trend, i.e., that both slopes and intercepts are independent random variables governed by common hyperparameters.

16.3.1 Analysis Using R

```
library('lme4')
glmm.fit  <- glmer(C ~ year + (1 | pop) + ( 0+ year | pop), family = poisson)
glmm.fit                   # Inspect results

> glmm.fit# Inspect results
Generalized linear mixed model fit by the Laplace approximation
Formula: C ~ year + (1 | pop) + (0 + year | pop)
    AIC    BIC   logLik deviance
  602.7 619.3  -297.3    594.7
Random effects:
  Groups Name        Variance    Std.Dev.
  pop    (Intercept) 0.67721     0.82293
  pop    year        0.13311     0.36484
Number of obs: 480, groups: pop, 16

Fixed effects:
              Estimate  Std. Error  z value  Pr(>|z|)
(Intercept)   3.2406      0.2071     15.65   <2e-16  ***
year         -1.8233      0.1055    -17.28   <2e-16  ***
- - -
Signif. codes: 0 '***' 0.001 '**' 0.01 '*' 0.05 '.' 0.1 ' ' 1

Correlation of Fixed Effects:
     (Intr)
year -0.044
```

16.3.2 Analysis Using WinBUGS

```
# Define model
sink("glmm.txt")
cat("
model {

# Priors
  for (i in 1:n.groups){
    alpha[i] ~ dnorm(mu.int, tau.int)      # Intercepts
    beta[i] ~ dnorm(mu.beta, tau.beta)     # Slopes
  }
  mu.int ~ dnorm(0, 0.001)     # Hyperparameter for random intercepts
  tau.int <- 1 / (sigma.int * sigma.int)
  sigma.int ~ dunif(0, 10)
```

```
  mu.beta ~ dnorm(0, 0.001)    # Hyperparameter for random slopes
  tau.beta <- 1 / (sigma.beta * sigma.beta)
  sigma.beta ~ dunif(0, 10)

# Poisson likelihood
  for (i in 1:n) {
    C[i] ~ dpois(lambda[i])
    lambda[i] <- exp(alpha[pop[i]] + beta[pop[i]]* year[i])
  }
}
",fill=TRUE)
sink()

# Bundle data
win.data <- list(C = C, pop = as.numeric(pop), year = year, n.groups = n.groups, n
= n)

# Inits function
inits <- function(){ list(alpha = rnorm(n.groups, 0, 2), beta = rnorm(n.groups, 0,
2), mu.int = rnorm(1, 0, 1), sigma.int = rlnorm(1), mu.beta = rnorm(1, 0, 1),
sigma.beta = rlnorm(1))}

# Parameters to estimate
parameters <- c("alpha", "beta", "mu.int", "sigma.int", "mu.beta", "sigma.beta")

# MCMC settings
ni <- 2000
nb <- 500
nt <- 2
nc <- 3

# Start Gibbs sampling
out <- bugs(win.data, inits, parameters, "glmm.txt", n.thin=nt, n.chains=nc,
n.burnin=nb, n.iter=ni, debug = TRUE)
```

This GLMM converges easily.

```
print(out, dig = 2)
> print(out, dig = 2)
Inference for Bugs model at "glmm.txt", fit using WinBUGS,
 3 chains, each with 2000 iterations (first 500 discarded), n.thin = 2
 n.sims = 2250 iterations saved
             mean   sd   2.5%    25%    50%    75%  97.5% Rhat n.eff
[ ... ]
mu.int       3.24 0.24   2.77   3.08   3.23   3.40   3.75 1.00  2200
sigma.int    0.94 0.20   0.64   0.79   0.91   1.04   1.44 1.01   280
mu.beta     -1.83 0.12  -2.08  -1.90  -1.82  -1.74  -1.60 1.00   870
sigma.beta   0.41 0.10   0.25   0.34   0.40   0.47   0.65 1.00   970
[ ... ]
```

```
DIC info (using the rule, pD = var(deviance)/2)
pD = 31.7 and DIC = 2478.9
DIC is an estimate of expected predictive error (lower deviance is better).
```

```
# Compare with input values
intercept.mean  ;  intercept.sd  ;  slope.mean  ;  slope.sd
> intercept.mean  ;  intercept.sd  ;  slope.mean  ;  slope.sd
[1] 3
[1] 1
[1] -2
[1] 0.6
>
```

This comparison looks good.

16.4 SUMMARY

We have seen how the concept of random effects, which we first met in Chapters 9 and 12 for the normal linear case, can be carried over fairly smoothly to the non-normal case. Both frequentist and Bayesian analyses are fairly straightforward, but the model fitted in WinBUGS appears more transparent.

EXERCISES

1. *Swiss hare data 1*: Fit the random-coefficient Poisson regression to the counts without regard to land use. Take a single count per site and year. Don't forget to account for variable area by inclusion of an offset.

2. *Swiss hare data 2*: Now add a separate pair of hyperparameters (for intercept and slope) for arable and grassland areas; this should definitely get you published somewhere really nice. You could also add other explanatory variables, such as an index for fox abundance if you had that. Do not forget that you are not modeling *hare abundance*, just the *expected hare counts*. Nobody really knows what this means, but expected counts presumably represent an unknown, and you hope, constant, proportion of the true number of hares out there.

CHAPTER

17

Binomial "t-Test"

17.1 INTRODUCTION

The next three chapters deal with another common kind of count data, where we want to estimate a binomial proportion. The associated models are often called logistic regressions. The crucial difference between binomial and Poisson random variables is the presence of a ceiling in the former: binomial counts cannot be greater than some upper limit. Therefore, we model bounded counts, where the bound is provided by trial size N. It makes sense to express a count relative to that bound, and this yields a proportion. In contrast, Poisson counts lack an upper limit, at least in principle.

Modeling a binomial random variable in essence means modeling a series of coin flips. We count the number of heads among the total number of coin flips (N) and from this want to estimate the general tendency of the coin to show heads. That is, we want to estimate Pr(heads). Data coming from coin flip-like processes are ubiquitous in nature and include survival or the occurrence of an organism. In coin flips, the binomial distribution describes the number of times r a coin shows heads among a number of N flips, where the coin has Pr(heads) = p. We also write $r \sim$ Binomial (N, p). A special case of the binomial distribution with $N = 1$, corresponding to

FIGURE 17.1   Cross-leaved gentian (*Gentiana cruciata*), Spanish Pyrenees, 2006. (*Photo* M. *Kéry*)

a single flip of the coin, is called the Bernoulli distribution. It has just a single parameter p.

As our inferential setting of this chapter, we consider a plant inventory on calcareous grasslands in the Jura mountains. A total of 50 sites were visited by experienced botanists who recorded whether they saw a species or not. The Cross-leaved gentian (Fig. 17.1) was found at 13 sites and the Chiltern gentian (see Chapter 20) at 29 sites. We wonder whether this is enough evidence, given the variation inherent in binomial sampling, to claim that the Cross-leaved gentian has a more restricted distribution in the Jura mountains.

This type of data is often called "presence–absence data." It is more accurate to call it "detection–nondetection data," since the number of sites at which a species is detected depends on two entirely different things: first, the number of sites where a species is actually present and second, the ease with which a species is detected at an occupied site. Without a special kind of data (see Chapter 20), we have no way of distinguishing between the two components of "presence–absence." All we can do is hope, pray, or claim either that both gentian species are found at every occupied site or else that their probability to be overlooked is identical. For now, we assume that every individual present is detected, i.e., that every occupied site is observed as such. Given such data, our question can be framed statistically by what can be called a binomial version of the t-test, i.e. a logistic regression that contrasts two groups. For gentian species i, let C_i be the number of sites it was detected. A simple model for C_i is this:

1. (Statistical) Distribution: $C_i \sim$ Binomial (N, p_i)

2. Link function: logit, i.e., $\text{logit}(p_i) = \log\left(\dfrac{p_i}{1 - p_i}\right) = $ linear predictor

3. Linear predictor: $\alpha + \beta * x_i$

If x is an indicator for the Chiltern gentian, then α can be interpreted as a logit-scale parameter for the probability of occurrence of the Cross-leaved gentian in the Jura mountains and β is the difference, on a logit-scale, between the probability of occurrence of the Chiltern gentian and that of the Cross-leaved gentian.

17.2 DATA GENERATION

We simulate the data from a binomial process whose parameters were defined such that the sample data approximately match those in our example. Note that our modeled response simply consists of two numbers.

```
N <- 50                      # Binomial total (Number of coin flips)
p.cr <- 13/50                # Success probability Cross-leaved
p.ch <- 29/50                # Success probability Chiltern gentian

C.cr <- rbinom(1, 50, prob = p.cr)   ;   C.cr      # Add Binomial noise
C.ch <- rbinom(1, 50, prob = p.ch)   ;   C.ch      # Add Binomial noise
C <- c(C.cr, C.ch)
species <- factor(c(0,1), labels = c("Cross-leaved gentian","Chiltern gentian"))
```

17.3 ANALYSIS USING R

A binomial t-test in R suggests a significant difference in the perceived distribution of the Cross-leaved gentian and the Chiltern gentian.

```
summary(glm(cbind(C, N - C) ~ species, family = binomial))
predict(glm(cbind(C, N - C) ~ species, family = binomial), type = "response")

> summary(glm(cbind(C, N - C) ~ species, family = binomial))

Call:
glm(formula = cbind(C, N - C) ~ species, family = binomial)
[ ... ]
```

```
Coefficients:
                        Estimate   Std. Error   z value   Pr(>|z|)
(Intercept)              -1.0460       0.3224    -3.244    0.00118  **
speciesChiltern gentian   1.3687       0.4313     3.173    0.00151  **
- - -
[ ... ]

> predict(glm(cbind(C, N - C) ~ species, family = binomial), type = "response")
    1    2
0.26 0.58
```

17.4 ANALYSIS USING WinBUGS

```
# Define model
sink("Binomial.t.test.txt")
cat("
model {

# Priors
 alpha ~ dnorm(0,0.01)
 beta ~ dnorm(0,0.01)

# Likelihood
 for (i in 1:n) {
    C[i] ~ dbin(p[i], N)        # Note p before N
    logit(p[i]) <- alpha + beta *species[i]
 }
# Derived quantities
 Occ.cross <- exp(alpha) / (1 + exp(alpha))
 Occ.chiltern <- exp(alpha + beta) / (1 + exp(alpha + beta))
 Occ.Diff <- Occ.chiltern - Occ.cross       # Test quantity
}
",fill=TRUE)
sink()

# Bundle data
win.data <- list(C = C, N = 50, species = c(0,1), n = length(C))

# Inits function
inits <- function(){ list(alpha=rlnorm(1), beta=rlnorm(1))}

# Parameters to estimate
params <- c("alpha", "beta", "Occ.cross", "Occ.chiltern", "Occ.Diff")
```

```
# MCMC settings
nc <- 3
ni <- 1200
nb <- 200
nt <- 2

# Start Gibbs sampling
out <- bugs(data=win.data, inits=inits, parameters.to.save=params,
model.file="Binomial.t.test.txt", n.thin=nt, n.chains=nc, n.burnin=nb, n.iter=ni,
debug = TRUE)

print(out, dig = 3)
> print(out, dig = 3)
Inference for Bugs model at "Binomial.t.test.txt", fit using WinBUGS,
 3 chains, each with 1200 iterations (first 200 discarded), n.thin = 2
 n.sims = 1500 iterations saved
                 mean     sd    2.5%     25%     50%     75%   97.5%  Rhat n.eff
alpha          -1.047  0.333  -1.710  -1.273  -1.034  -0.810  -0.441 1.007   290
beta            1.362  0.449   0.520   1.059   1.375   1.657   2.278 1.007   300
Occ.cross       0.265  0.063   0.153   0.219   0.262   0.308   0.392 1.007   290
Occ.chiltern    0.577  0.069   0.438   0.530   0.580   0.626   0.701 1.003   960
Occ.Diff        0.312  0.095   0.124   0.251   0.318   0.376   0.485 1.006   340
[ ... ]
DIC info (using the rule, pD = Dbar-Dhat)
pD = 2.1 and DIC = 12.6
```

Next, we plot the posterior distribution of the (biological) distributions of the two species and their difference, as perceived from the detection–nondetection data (Fig. 17.2):

```
par(mfrow = c(3,1))
hist(out$sims.list$Occ.cross, col = "grey", xlim = c(0,1), main = "", xlab =
"Occupancy Cross-leaved", breaks = 30)
abline(v = out$mean$Occ.cross, lwd = 3, col = "red")
hist(out$sims.list$Occ.chiltern, col = "grey", xlim = c(0,1), main = "", xlab =
"Occupancy Chiltern", breaks = 30)
abline(v = out$mean$Occ.chiltern, lwd = 3, col = "red")
hist(out$sims.list$Occ.Diff, col = "grey", xlim = c(0,1), main = "", xlab =
"Difference in occupancy", breaks = 30)
abline(v = 0, lwd = 3, col = "red")
```

The posterior distribution of the difference barely overlaps zero (Fig. 17.2, bottom), and the 95% credible interval is (0.015, 0.485). Therefore, we can say that the Chiltern gentian is found at significantly more sites than the Cross-leaved gentian. This is not equivalent to saying that the Chiltern gentian was more widespread in the Jura mountains than is

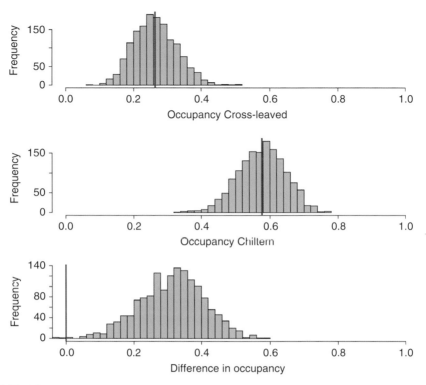

FIGURE 17.2 Posterior distributions of the occupancy probability of Cross-leaved (top panel) and Chiltern gentian (middle panel); vertical red lines indicate posterior means. Bottom panel shows the posterior distribution of the difference of the two species' occupancy probability; vertical red line indicates zero.

the Cross-leaved gentian, only that it is *detected* at more sites. Whether it also means the former depends on whether the assumption of perfect or at least of constant detection probability holds (MacKenzie and Kendall, 2002; Kéry and Schmidt, 2008).

17.5 SUMMARY

As for all generalized linear models, a binomial t-test in WinBUGS is a fairly trivial generalization of the corresponding normal response model. We have seen another example of the ease with which, in Markov chain Monte Carlo-based statistical inference, we can compute derived parameters exactly, i.e., without any approximations, and with full error propagation.

EXERCISES

1. *Binomial and Bernoulli*: Try to formulate the same problem using a Bernoulli distribution. You will need to simulate, not the aggregate number of sites at which each gentian species was found, but rather the detected occupancy status of each individual site. Adapt the code for analysis. In WinBUGS, you can directly specify a Bernoulli by `dbern(p)`. (As usual, when you're unsure, go to the WinBUGS manual: `Help > User Manual` and then scroll down.)

2. *Swiss hare data*: Model the probability of occurrence of a "large" count in arable and grassland areas. Select a useful threshold to call a count "large." Are large counts more common in arable areas?

Binomial Analysis of Covariance

18.1 INTRODUCTION

We can specify a binomial analysis of covariance (ANCOVA) by adding discrete and continuous covariates to the linear predictor of a binomial generalized linear model (GLM). Once again, to stress the structural similarity with the normal linear model in Chapter 11, we modify the asp viper example just slightly. Instead of modeling a continuous measurement such as body mass in Chapter 11, we model a count governed by an underlying probability; specifically, we model the proportion of black individuals in adder populations. The adder has an all-black and a zigzag morph, where females are brown and males are gray (Fig. 18.1).

It has been hypothesized that the black color confers a thermal advantage, and therefore, the proportion of black individuals should be greater in cooler or wetter habitats. We simulate data that bear on this question and "study," by simulation, 10 adder populations each in the Jura mountains, the Black Forest, and the Alps. We capture a number of snakes in these populations and record the proportion of black adders. Then, we relate these proportions to the mountain range and to a combined index of low

Introduction to WinBUGS for Ecologists
DOI: 10.1016/B978-0-12-378605-0.00018-1

FIGURE 18.1   Male adder (*Vipera berus*) of the zigzag morph, Germany, 2007. (*Photo T. Ott*)

temperature, wetness, and northerliness of the site. Our expectation will of course be that there are relatively more black adders at cool and wet sites. As always, a count is the result of a true number (here, of black and zigzag adders) and a detection probability. Hence, in the following analyses, we make the implicit assumption that the detectability of black and zigzag adders neither differs between each other nor among populations.

We model the number of black adders C_i among N_i captured animals in population i. Here is the description of the model.

1. Distribution: $C_i \sim \text{Binomial}(p_i, N_i)$

2. Link function: logit, i.e., $\text{logit}(p_i) = \log\left(\dfrac{p_i}{1 - p_i}\right) = $ linear predictor

3. Linear predictor: $\alpha_{\text{Jura}} + \beta_1 * x_{\text{BlackF}} + \beta_2 * x_{\text{Alps}} + \beta_3 * x_{\text{wet}} + \beta_4 * x_{\text{wet}} * x_{\text{BlackF}} + \beta_5 * x_{\text{wet}} * x_{\text{Alps}}$

Note that the number of animals captured, N_i, is not a parameter of the binomial distribution in this example but is known (in contrast to the model of Chapter 21 where N_i will be estimated). The link function is the logit, as is customary for a binomial distribution, though other links are possible (e.g., the complementary log–log, which is asymmetrical; see GLM textbooks such as McCullagh and Nelder, 1989). Finally, the linear predictor is made up of an intercept α_{Jura} for the expected proportion, on the logit scale, of black adders in the Jura and parameters β_1 and β_2 for the difference in the intercept between Black Forest and the Jura and the Alps and the Jura, respectively. The parameters β_3, β_4, and β_5 specify the slope of the (logit-linear) relationship between the proportion of black adders and the wetness indicator of a site in the Jura and the difference from β_3 of these slopes in the Black Forest and the Alps, respectively.

18.2 DATA GENERATION

```
n.groups <- 3
n.sample <- 10
n <- n.groups * n.sample
x <- rep(1:n.groups, rep(n.sample, n.groups))
pop <- factor(x, labels = c("Jura", "Black Forest", "Alps"))
```

We construct a continuous wetness index: 0 denotes wet sites lacking sun and 1 is the converse. For ease of presentation, we sort this covariate; this has no effect on the analysis.

```
wetness.Jura <- sort(runif(n.sample, 0, 1))
wetness.BlackF <- sort(runif(n.sample, 0, 1))
wetness.Alps <- sort(runif(n.sample, 0, 1))
wetness <- c(wetness.Jura, wetness.BlackF, wetness.Alps)
```

We also need the number of adders examined in each population (N_i), i.e., the binomial totals, also called sample or trial size of the binomial distribution. We assume that the total number of snakes examined in each population is a random variable drawn from a uniform distribution, but this is not essential.

```
N <- round(runif(n, 10, 50) )        # Get discrete Uniform values
```

We build the design matrix of an interactive combination of population and wetness.

```
Xmat <- model.matrix(~ pop*wetness)
print(Xmat, dig = 2)
```

Select the parameter values, i.e., choose values for α_{Jura}, β_1, β_2, β_3, β_4, and β_5.

```
beta.vec <- c(-4, 1, 2, 6, 2, -5)
```

We assemble the number of black adders captured in each population in three steps:

1. We add up all components of the linear model to get the value of the linear predictor,

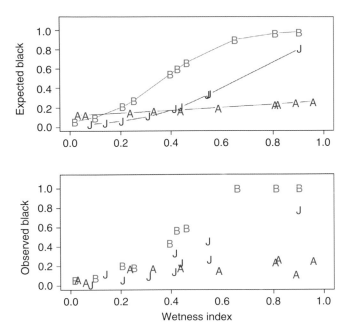

FIGURE 18.2 Relationship between the wetness index and the expected (top) and observed proportion (bottom) of black adders in the Jura mountains (red J), the Black Forest (green B), and the Alps (blue A). The discrepancy between the two graphs is due to binomial sampling variation.

2. We apply the inverse logit transformation to get the expected proportion (p_i) of black adders in each population (Fig. 18.2; top), and finally,
3. We add binomial noise, i.e., use p_i and N_i to draw binomial random numbers representing the count of black adders in each sample of N_i snakes (Fig. 18.2; bottom).

The value of the linear predictor is again obtained by matrix multiplication of the design matrix (Xmat) and the parameter vector (beta.vec). (For ecological purists the usual reality disclaimer applies again.).

```
lin.pred <- Xmat[,] %*% beta.vec              # Value of lin.predictor
exp.p <- exp(lin.pred) / (1 + exp(lin.pred))  # Expected proportion

C <- rbinom(n = n, size = N, prob = exp.p)    # Add binomial noise
hist(C)                                        # Inspect what we've created

par(mfrow = c(2,1))
matplot(cbind(wetness[1:10], wetness[11:20], wetness[21:30]), cbind(exp.p[1:10],
exp.p[11:20], exp.p[21:30]), ylab = "Expected black", xlab = "", col =
c("red","green","blue"), pch = c("J","B","A"), lty = "solid", type = "b", las = 1,
cex = 1.2, main = "", lwd = 1)
```

```
matplot(cbind(wetness[1:10], wetness[11:20], wetness[21:30]),
cbind(C[1:10]/N[1:10], C[11:20]/N[11:20], C[21:30]/N[21:30]), ylab = "Observed
black", xlab = "Wetness index", col = c("red","green","blue"), pch =
c("J","B","A"), las = 1, cex = 1.2, main = "")
par(mfrow = c(1,1))
```

Hence, the proportion of black adders increases with wetness most steeply in the Black Forest, less so in the Jura, and hardly at all in the Alps (Fig. 18.2).

18.3 ANALYSIS USING R

As usual, the analysis in R is concise.

```
summary(glm(cbind(C, N-C) ~ pop * wetness, family = binomial))
beta.vec

> summary(glm(cbind(C, N-C) ~ pop * wetness, family = binomial))

Call:
glm(formula = cbind(C, N-C) ~ pop * wetness, family = binomial)

Deviance Residuals:
    Min         1Q     Median        3Q        Max
 -1.8911    -0.5785     0.1309    0.8162     1.8866

Coefficients:
                        Estimate   Std. Error   z value   Pr(>|z|)
(Intercept)             -3.7255       0.3905     -9.541   < 2e-16 ***
popBlack Forest          0.1342       0.5894      0.228   0.81990
popAlps                  1.5118       0.4946      3.057   0.00224 **
wetness                  5.5286       0.7215      7.663   1.81e-14 ***
popBlack Forest:wetness  3.6046       1.3070      2.758   0.00582 **
popAlps:wetness         -4.3748       0.8607     -5.083   3.72e-07 ***
- - -
Signif. codes: 0 '***' 0.001 '**' 0.01 '*' 0.05 '.' 0.1 ' ' 1

(Dispersion parameter for binomial family taken to be 1)

    Null deviance: 374.63  on 29  degrees of freedom
Residual deviance:  26.41  on 24  degrees of freedom
AIC: 123.75

Number of Fisher Scoring iterations: 4

> print(beta.vec)
[1] -4  1   2  6  2  -5
>
```

Owing to the small sample size, we observe only a moderate correspondence with the input values.

18.4 ANALYSIS USING WinBUGS

In the following WinBUGS code, we add a few lines to compute Pearson residuals. For a binomial response, these can be computed as $(C_i - N_i p_i) / \sqrt{N_i p_i (1 - p_i)}$, where C_i is the binomial count for unit i and N_i and p_i are the trial size and success probability of the associated binomial distributions. The Pearson residual has the form of a raw residual divided by the standard deviation of unit i.

```
# Define model
sink("glm.txt")
cat("
model {

# Priors
 for (i in 1:n.groups){
    alpha[i] ~ dnorm(0, 0.01)        # Intercepts
    beta[i] ~ dnorm(0, 0.01)         # Slopes
 }

# Likelihood
 for (i in 1:n) {
    C[i] ~ dbin(p[i], N[i])
    logit(p[i]) <- alpha[pop[i]] + beta[pop[i]]* wetness[i] # Baseline Jura

# Fit assessments: Pearson residuals and posterior predictive check
    Presi[i] <- (C[i]-N[i]*p[i]) / sqrt(N[i]*p[i]*(1-p[i])) # Pearson resi
    C.new[i] ~ dbin(p[i], N[i])                 # Create replicate data set
    Presi.new[i] <- (C.new[i]-N[i]*p[i]) / sqrt(N[i]*p[i]*(1-p[i]))
    D[i] <- pow(Presi[i], 2)                    # Squared Pearson residuals
    D.new[i] <- pow(Presi.new[i], 2)
 }

# Derived quantities
# Recover the effects relative to baseline level (no. 1)
 a.effe2 <- alpha[2] - alpha[1]        # Intercept Black Forest vs. Jura
 a.effe3 <- alpha[3] - alpha[1]        # Intercept Alps vs. Jura
 b.effe2 <- beta[2] - beta[1]          # Slope Black Forest vs. Jura
 b.effe3 <- beta[3] - beta[1]          # Slope Alps vs. Jura

# Custom tests
 test1 <- beta[3] - beta[2]            # Difference slope Alps -Black Forest

# Add up discrepancy measures
 fit <- sum(D[])
 fit.new <- sum(D.new[])
 }
```

```
",fill=TRUE)
sink()

# Bundle data
win.data <- list(C = C, N = N, pop = as.numeric(pop), n.groups = n.groups, wetness
= wetness, n = n)

# Inits function
inits <- function(){ list(alpha = rlnorm(n.groups, 3, 1), beta = rlnorm(n.groups,
2, 1))}             # Note log-normal inits here

# Parameters to estimate
params <- c("alpha", "beta", "a.effe2", "a.effe3", "b.effe2", "b.effe3", "test1",
"Presi", "fit", "fit.new")

# MCMC settings
ni <- 1500
nb <- 500
nt <- 5
nc <- 3

# Start Gibbs sampler
out <- bugs(data = win.data, inits = inits, parameters.to.save = params, model.file
= "glm.txt", n.thin = nt, n.chains = nc, n.burnin = nb, n.iter = ni, debug = TRUE)
```

The model converges rapidly. To practice good statistical behavior, we now first assess model fit before rushing on to inspect the parameter estimates (or indulge in some activity related to significance testing). After all, only when a model adequately reproduces the salient features in the data set should we believe what it says about the parameters. We first plot the Pearson residuals and, naturally, find no obvious remaining pattern related to order or wetness of a site (Fig. 18.3).

```
par(mfrow = c(1,2), cex = 1.5)
plot(out$mean$Presi, ylab = "Residual", las = 1)
abline(h = 0)
plot(wetness, out$mean$Presi, ylab = "Residual", las = 1)
abline(h = 0)
```

Next, we conduct a posterior predictive check for overall goodness of fit of the model (Fig. 18.4). Remember that our discrepancy measure is the sum of squared Pearson residuals.

```
plot(out$sims.list$fit, out$sims.list$fit.new, main = "", xlab = "Discrepancy
actual data", ylab = "Discrepancy ideal data")
abline(0,1, lwd = 2, col = "black")
```

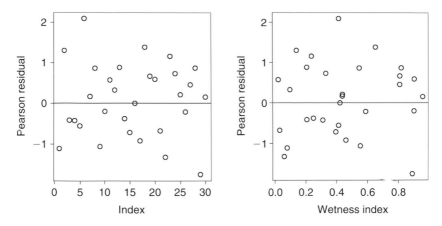

FIGURE 18.3 Pearson residuals plotted against the order of each observation (left) and against the value of the wetness indicator (right).

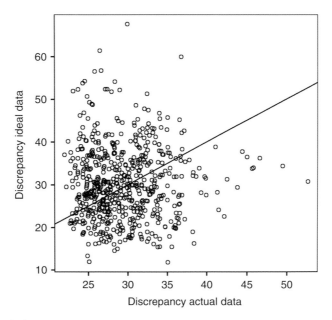

FIGURE 18.4 Posterior predictive check for the black adder analysis. The Bayesian p-value (here, 0.52) is the proportion of points above the line.

Of course, this looks good and so does the Bayesian *p*-value.

```
mean(out$sims.list$fit.new > out$sims.list$fit)
> mean(out$sims.list$fit.new > out$sims.list$fit)
[1] 0.52
```

Hence, we feel justified in comparing the results of the Bayesian analysis with the truth from the data-generating random process and with the frequentist inference from R's function glm( ).

```
print(out, dig = 3)                    # Bayesian analysis
beta.vec                               # Truth in data generation
print(glm(cbind(C, N-C) ~ pop * wetness, family = binomial)$coef, dig = 4)
                                       # The ML solution
```

We get rather similar estimates. Parameters in the output of the Bayesian analysis printed in boldface correspond to the quantities used for data generation and also to those of the model parameterization in the frequentist analysis in R.

```
> print(out, dig = 3)                         # Bayesian analysis
Inference for Bugs model at "glm.txt", fit using WinBUGS,
 3 chains, each with 1500 iterations (first 500 discarded), n.thin = 5
 n.sims = 600 iterations saved
```

	mean	sd	2.5%	25%	50%	75%	97.5%	Rhat	n.eff
alpha[1]	**-3.797**	0.383	-4.514	-4.056	-3.782	-3.518	-3.048	1.009	190
alpha[2]	-3.591	0.421	-4.408	-3.896	-3.580	-3.302	-2.825	1.002	600
alpha[3]	-2.225	0.301	-2.837	-2.422	-2.235	-2.021	-1.651	1.031	68
beta[1]	**5.655**	0.701	4.372	5.164	5.632	6.131	7.004	1.009	200
beta[2]	9.139	1.086	7.148	8.362	9.102	9.927	11.300	1.000	600
beta[3]	1.162	0.477	0.180	0.843	1.192	1.476	2.086	1.034	60
a.effe2	0.207	0.597	-0.972	-0.190	0.201	0.631	1.289	1.012	200
a.effe3	1.572	0.480	0.639	1.256	1.559	1.894	2.536	1.018	110
b.effe2	3.484	1.337	0.960	2.543	3.490	4.472	6.148	1.009	360
b.effe3	-4.492	0.832	-6.134	-5.000	-4.471	-3.916	-3.022	1.012	150
test1	-7.977	1.164	-10.501	-8.788	-7.897	-7.218	-5.741	1.007	280
Presi[1]	-1.109	0.185	-1.520	-1.235	-1.103	-0.979	-0.805	1.009	190

```
[ ... ]
DIC info (using the rule, pD = Dbar-Dhat)
pD = 6.0 and DIC = 123.8
DIC is an estimate of expected predictive error (lower deviance is better).

> beta.vec                          # Truth in data-generation
[1] -4  1  2  6  2 -5
```

```
> print(glm(cbind(C, N-C) ~ pop * wetness, family = binomial)$coef, dig = 4)
                                   # The ML solution
        (Intercept)         popBlack Forest              popAlps
            -3.7255                  0.1342               1.5118
            wetness  popBlack Forest:wetness      popAlps:wetness
             5.5286                  3.6046              -4.3748
```

18.5 SUMMARY

Moving from the normal and the Poisson to a binomial ANCOVA involves only minor changes in the code of WinBUGS (and also R). Similarly, the concepts of residuals and posterior predictive distributions carry over to this class of models. We see examples for both.

EXERCISES

1. *Multiple binomial regression*: Create another habitat variable X and, using the data set created in this chapter, use WinBUGS to fit the following model prop.black ~ pop*wet + X + wet:X. That is, add the main effect of X plus the interaction between wet and X.

2. *Swiss hare data*: Model the probability of a "large" count as a function of land use and year (continuous), i.e., fit the following model:

   large.pop ~ land-use * year

   You may choose a threshold of about 5. You may also want to check for nonlinearity (on the scale of the logit link) by adding a quadratic effect of year.

Binomial Mixed-Effects Model (Binomial GLMM)

19.1 INTRODUCTION

As in a Poisson generalized linear mixed model (GLMM), we can also add into a binomial generalized linear model (GLM) random variation beyond what is stipulated by the binomial distribution. We show this for a slight modification of the Red-backed shrike example from Chapter 16. Instead of counting the number of pairs, which naturally leads to the adoption of a Poisson model, we now study the reproductive success (success or failure) of its much rarer cousin, the woodchat shrike (Fig. 19.1). We examine the relationship between precipitation during the breeding season and reproductive success; wet springs are likely to depress the proportion of successful nests. We assemble data from 16 populations studied over 10 years.

First, we write down the random-coefficients model (without intercept-slope correlation) for a binomial response. We model C_i, the number of successful pairs among N_i studied pairs in year i and study area j:

1. Distribution: $C_i \sim \text{Binomial}(p_i, N_i)$
2. Link function: logit, i.e., $\text{logit}(p_i) = \log\left(\dfrac{p_i}{1 - p_i}\right) = $ linear predictor

FIGURE 19.1   Woodchat shrike (*Lanius senator*), Catalonia, 2008. (*Photo J. Rojals*)

3. Linear predictor: $\alpha_{j(i)} + \beta_{j(i)} * x_i$
4. Submodel for parameters: $\alpha_j \sim \text{Normal}(\mu_\alpha, \sigma_\alpha^2)$
$$\beta_j \sim \text{Normal}(\mu_\beta, \sigma_\beta^2)$$

Except for a different distribution and link function, the additional kind of data provided by the binomial totals N_i, and a different interpretation and indexing of the covariate, this model looks exactly like the Poisson GLMM in Chapter 16! The linear predictor, $\alpha_{j(i)} + \beta_{j(i)} * x_i$, specifies a population-specific, logit-linear relationship between breeding success and precipitation. Furthermore, populations are assumed to be related in the sense that both intercepts (α_j) and slopes (β_j) of those relationships come from two normal distributions whose hyperparameters we estimate.

19.2 DATA GENERATION

We generate data under the random-coefficients model, i.e., with both α_j and β_j assumed independent and random effects. We assume no correlation.

```
n.groups <- 16
n.years <- 10
n <- n.groups * n.years
pop <- gl(n = n.groups, k = n.years)
```

We create a uniform covariate as an index to spring precipitation: 0 denotes little rain and 1 much.

```
prec <- runif(n, 0, 1)
```

N_i, the binomial total, is the number of nesting attempts surveyed in year i.

```
N <- round(runif(n, 10, 50) )
```

We build the design matrix as before.

```
Xmat <- model.matrix(~pop*prec-1-prec)
print(Xmat[1:91,], dig = 2)              # Print top 91 rows
```

Next, we choose the parameter values from their respective normal distributions, but first, we need to select the associated hyperparameters.

```
intercept.mean <- 1        # Select hyperparams
intercept.sd <- 1
slope.mean <- -2
slope.sd <- 1
intercept.effects<-rnorm(n = n.groups, mean = intercept.mean, sd = intercept.sd)
slope.effects <- rnorm(n = n.groups, mean = slope.mean, sd = slope.sd)
all.effects <- c(intercept.effects, slope.effects)   # Put them all together
```

We assemble the counts C_i by first computing the value of the linear predictor, then applying the inverse-logit transformation, and finally integrating binomial noise (where we need N_i).

```
lin.pred <- Xmat %*% all.effects            # Value of lin.predictor
exp.p <- exp(lin.pred) / (1 + exp(lin.pred))    # Expected proportion
```

For each population, we plot the expected (Fig. 19.2) and the observed breeding success (Fig. 19.3) of woodchat shrikes against standardized spring precipitation using a Trellis plot.

```
library("lattice")
xyplot(exp.p ~ prec | pop, ylab = "Expected woodchat shrike breeding success ",
xlab = "Spring precipitation index", main = "Expected breeding success")

C <- rbinom(n = n, size = N, prob = exp.p)        # Add binomial variation
xyplot(C/N ~ prec | pop, ylab = "Realized woodchat shrike breeding success", xlab
= "Spring precipitation index", main = "Realized breeding success")
```

19.3 ANALYSIS UNDER
A RANDOM-COEFFICIENTS MODEL

We could assume that all shrike populations have the same relationship between breeding success and standardized spring precipitation, but at different levels, corresponding to a random-intercepts model

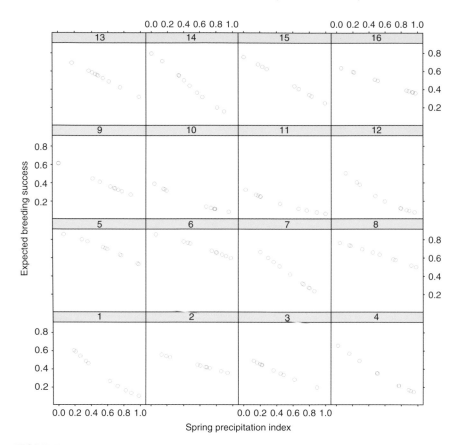

FIGURE 19.2   Trellis plot of the relationship between spring precipitation (standardized) and expected breeding success of woodchat shrikes in 16 populations over 10 years.

(see Section 12.3). However, we directly adopt the random coefficients model without correlation instead. This means assuming that every shrike population has a specific response to precipitation but that both intercept and slope are "similar" among populations.

19.3.1 Analysis Using R

```
library('lme4')
glmm.fit <- glmer(cbind(C, N-C) ~ prec + (1 | pop) + ( 0+ prec | pop), family =
binomial)
glmm.fit
```

```
> glmm.fit
Generalized linear mixed model fit by the Laplace approximation
Formula: cbind(C, N - C) ~ prec + (1 | pop) + (0 + prec | pop)
```

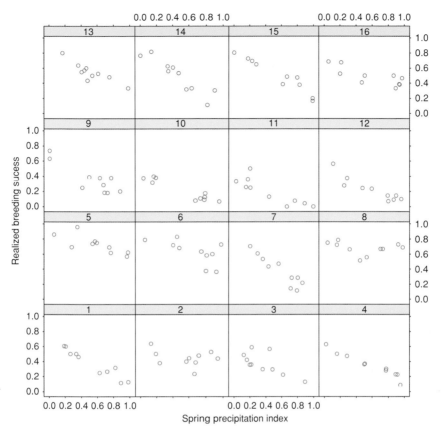

FIGURE 19.3   Trellis plot of the relationship between spring precipitation (standardized) and the realized breeding success of woodchat shrikes in 16 populations over 10 years. The difference between this and the previous plot is due to binomial sampling variation.

```
      AIC    BIC  logLik  deviance
    229.3  241.6  −110.6     221.3
Random effects:
 Groups Name         Variance  Std.Dev.
 pop    (Intercept)  0.38785   0.62277
 pop    prec         0.53258   0.72978
Number of obs: 160, groups: pop, 16

Fixed effects:
             Estimate  Std.Error  z value  Pr(>|z|)
(Intercept)    0.8006     0.1690    4.738  2.16e-06 ***
prec          −2.1290     0.2173   −9.796   < 2e-16 ***
---
Signif. codes: 0 '***' 0.001 '**' 0.01 '*' 0.05 '.' 0.1 ' ' 1
```

```
Correlation of Fixed Effects:
      (Intr)
prec -0.182
```

19.3.2 Analysis Using WinBUGS

```
# Define model
sink("glmm.txt")
cat("
model {

# Priors
 for (i in 1:n.groups){
    alpha[i] ~ dnorm(mu.int, tau.int)     # Intercepts
    beta[i] ~ dnorm(mu.beta, tau.beta)    # Slopes
 }
 mu.int ~ dnorm(0, 0.001)     # Hyperparameter for random intercepts
 tau.int <- 1 / (sigma.int * sigma.int)
 sigma.int ~ dunif(0, 10)

 mu.beta ~ dnorm(0, 0.001)     # Hyperparameter for random slopes
 tau.beta <- 1 / (sigma.beta * sigma.beta)
 sigma.beta ~ dunif(0, 10)

# Binomial likelihood
 for (i in 1:n) {
    C[i] ~ dbin(p[i], N[i])
    logit(p[i]) <- alpha[pop[i]] + beta[pop[i]]* prec[i]
 }
}
",fill=TRUE)
sink()

# Bundle data
win.data <- list(C = C, N = N, pop = as.numeric(pop), prec = prec, n.groups =
n.groups, n = n)

# Inits function
inits <- function(){ list(alpha = rnorm(n.groups, 0, 2), beta = rnorm(n.groups, 1,
1), mu.int = rnorm(1, 0, 1), mu.beta = rnorm(1, 0, 1))}

# Parameters to estimate
params <- c("alpha", "beta", "mu.int", "sigma.int", "mu.beta", "sigma.beta")
```

```
# MCMC settings
ni <- 2000
nb <- 500
nt <- 2
nc <- 3
```

```
# Start Gibbs sampling
out <- bugs(win.data, inits, params, "glmm.txt", n.thin=nt, n.chains=nc,
n.burnin=nb, n.iter=ni, debug = TRUE)
```

This standard GLMM converges nicely.

```
print(out, dig = 2)
> print(out, dig = 2)
Inference for Bugs model at "glmm.txt", fit using WinBUGS,
 3 chains, each with 2000 iterations (first 500 discarded), n.thin = 2
 n.sims = 2250 iterations saved
              mean    sd   2.5%    25%    50%    75%  97.5%  Rhat n.eff
[ ... ]
mu.int        0.80  0.19   0.44   0.68   0.80   0.92   1.17  1.00  2200
sigma.int     0.70  0.16   0.45   0.59   0.68   0.79   1.08  1.00  1600
mu.beta      -2.13  0.25  -2.64  -2.28  -2.12  -1.97  -1.64  1.00  2200
sigma.beta    0.84  0.23   0.48   0.67   0.81   0.96   1.38  1.00  1100
[ ... ]
DIC info (using the rule, pD = var(deviance)/2)
pD = 32.8 and DIC = 748.4
DIC is an estimate of expected predictive error (lower deviance is better).
>
```

```
# Compare with input values
intercept.mean  ;  intercept.sd  ;  slope.mean  ;  slope.sd
> intercept.mean  ;  intercept.sd  ;  slope.mean  ;  slope.sd
[1] 1
[1] 1
[1] -2
[1] 1
>
```

This seems to work, too. As is typical, the estimates of random-effects variances are greater in the Bayesian approach, presumably since inference is exact and incorporates all uncertainty in the modeled system. In contrast, the approach in lmer() or glmer() is approximate and may underestimate these variances (Gelman and Hill, 2007).

19.4 SUMMARY

As for the Poisson case, the introduction of random effects into a binomial GLM in WinBUGS is particularly straightforward and transparent. Fitting the resulting binomial GLMM in WinBUGS is very helpful for your general understanding of this class of models.

EXERCISES

1. *Predictions*: Produce a plot of the mean expected response of breeding success to the spring precipitation index. Try to overlay these estimates on the observed data.

2. *Fixed and random, shrinkage*: Fit a fixed-effects binomial analysis of covariance (ANCOVA) model to the data set we just assembled and compare the estimates of the population-specific intercepts and slopes under the fixed- and the random-effects models.

20

Nonstandard GLMMs 1: Site-Occupancy Species Distribution Model

20.1 INTRODUCTION

We have now seen a wide range of random-effects (also called mixed or hierarchical) models, including the normal, Poisson, and binomial generalized linear models (GLMs) with random effects. This means that some parameters are themselves represented as realizations from a random process. With the exception of the zero-inflated models in chapter 14, we have used a normal (or a multivariate normal) distribution as our sole description of this additional random component. However, nothing constrains us to use the normal distribution only and sometimes, other distributions will be appropriate for some parameters. The next two chapters illustrate two cases with discrete random effects that are assumed to be drawn from a Bernoulli or a Poisson distribution. Importantly, these effects have a precise biological meaning in these models: they correspond to the true, but imperfectly observed, state of occurrence (Chapter 20) or of abundance (Chapter 21).

The modeling of animal and plant distributions is an important and active area of ecological research and applications. The most frequently applied method is a binomial GLM, or logistic regression (see Chapters 17–19), where the probability that an organism is found is modeled from what typically, and misleadingly, are called "presence–absence" data. These are binary indicators for whether a species was found (1) or not (0) in a spatial sample unit, and the effect of covariates is modeled through the logit link function. There are many variants of this approach, but the basic principle is often the same: so-called "presence–absence" data are directly modeled as coming from a Bernoulli distribution and the Bernoulli parameter is interpreted as the probability of occurrence.

However, a fundamental and extremely widely overlooked issue in almost all species distribution models is that detectability (p) of most species is imperfect—typically, a species will not always be detected where it occurs (MacKenzie and Kendall, 2002; Pellet and Schmidt, 2005; Kéry and Schmidt, 2008; Kéry et al., 2010b). In other words, detection probability is typically less than one ($p < 1$). This basic fact is very well known to field naturalists and reasonably understood for animals, and it even applies to populations of immobile organisms such as plants (Kéry, 2004; Kéry et al., 2006). However, it seems to have been overlooked by most professional ecologists dealing with distributional data (Araujo and Guisan, 2006; Elith et al., 2006).

As a consequence, virtually no study generated by current species distribution modelers actually models the true occurrence of a species as pretended or believed. Rather, the product of the two probabilities of occurrence and detection of a species is modeled. For noteworthy exceptions, see Gelfand et al. (2005), Royle et al. (2005), Latimer et al. (2006), and Altwegg et al. (2008), also see Royle et al. (2007), and Webster et al. (2008).

The widespread confusion about what is actually being modeled in most species distribution models has three main consequences (MacKenzie et al., 2002; Tyre et al., 2003; Gu and Swihart, 2004; MacKenzie, 2006; MacKenzie et al., 2006; Kéry and Schmidt, 2008; Royle and Dorazio, 2008; Kéry et al., 2010b):

1. species distributions will be underestimated whenever $p < 1$,
2. estimates of covariate relationships will be biased toward zero whenever $p < 1$, and
3. the factors that affect the difficulty with which a species is found may end up in predictive models of species occurrence.

The first is easy to understand, but the second has not been widely recognized, although Tyre et al. (2003), Gu and Swihart (2004), and MacKenzie et al. (2006, pp. 34–35) described this effect already a few years ago. Interestingly, and in contrast to a naïve analysis of count data in the presence of imperfect detection (cf. simple Poisson regression examples in Chapters 13–16), in distribution modeling, even a constant $p < 1$ will bias low the slope estimate of the relationship between occurrence probability and a covariate. Presumably, this includes also a time covariate, i.e., situations where changes in distribution over time are modeled. As an example for the third effect, if a species has a higher probability to be found near roads, perhaps, because near roads, more people are likely to stumble upon it, then obviously roads or habitat types associated with roads will show up as important for that species, unless detection probability is accounted for. As an extreme example, when a species distribution map is constructed from road-kill records, then no matter how much roads might actually be avoided by that species in reality, the resulting distribution map will emphasize the great positive effect of roads on the distribution of the species!

In contrast, a novel class of models with the rather peculiar name "site-occupancy models" (MacKenzie et al., 2002, 2003, 2006; Tyre et al. 2003) is able to estimate the true distribution of animals or plants free of any distorting effects of the difficulty with which they are found. This chapter deals with these models and shows how to fit them using WinBUGS.

As a motivating example, we consider an inventory of the beautiful Chiltern gentian (Fig. 20.1) conducted in 150 calcareous grasslands in

FIGURE 20.1 Chiltern gentian (*Gentianella germanica*), Slovenia, 2007. (*Photo: M. Vogrin*)

the Jura mountains. Our aim is to estimate the proportion or number of occupied sites and to identify environmental factors related to the occurrence of the gentian. Interestingly, *Gentianella germanica* is a typical plant of nutrient-poor sites, which are thus *a priori* often rather dry. However, within the class of nutrient-poor grasslands, it preferentially occurs on wetter sites. However, these sites often have a higher and denser vegetation cover, so the rather small gentian (5–40 cm height) may more frequently be overlooked at these better sites (we ignore here the fact that better sites may hold larger populations). This effect could mask its preference for wetter sites. None of the currently widespread methods for distribution modeling such as GLM, generalized additive models (GAM), or boosted regression trees (Elith et al., 2006) are able to tease apart the effects of a covariate that influences both the occurrence of a species and the ease with which it is found (i.e. detection probability).

We will use site-occupancy models (MacKenzie et al., 2002, 2003, 2006) to separately estimate the gentian's probability of occurrence (called occupancy or species distribution) and its probability to be detected at occupied sites (detection probability), along with covariate effects on either occurrence or detection. It may be claimed that site-occupancy models are currently the only genuine distribution models available. All other widespread distribution modeling approaches confound occurrence and detection and only estimate the *apparent occurrence*, or more explicitly, the *combination of the probability of occurrence and the probability of detection, given the occurrence.*

The price to be paid for this improved inference is a sort of repeated-measures design, i.e., at least some sites need to be visited twice or preferably more frequently. This field protocol may be called a metapopulation design because the same quantity (occurrence) is assessed at many spatial replicates (Royle, 2004b; Royle and Dorazio, 2006). It is from the pattern of detection or nondetection at multiply visited sites that we obtain the information about detection probability, separate from occurrence probability. See MacKenzie and Royle (2005), MacKenzie et al. (2006), and Bailey et al. (2007) for design considerations relevant to this model.

A balanced design, i.e., an equal number of visits to all sites, is not essential for site-occupancy models; it simply makes things easier to simulate and present. (See Chapter 21 for an alternative, "vertical" data format, which is more convenient for the analysis of unbalanced metapopulation data.) Therefore, in our inventory of *G. germanica*, we assume that each site is visited three times by an independent botanist, and every time she notes whether at least one plant of *G. germanica* is detected or not. The result of these surveys may be summarized in a binary string, such as 010 for a site, where a gentian is detected during the second, but not during the first or third surveys. Generally, for a species surveyed T times at each of R sites, survey results are summarized in an R-by-T matrix containing a 1 when the species is detected at site i during survey j and a 0 when it is not.

The genesis, and therefore the analysis, of detection/nondetection observation y_{ij} at site i during survey j is naturally described by a hierarchical, or state-space, model that contains one submodel for the only partially observed true state (occurrence, the result of the biological process), and another submodel for the actual observations. The actual observations result from both the particular realization of the biological process and of the observation process.

$$z_i \sim \text{Bernoulli}(\psi) \qquad \text{Biological process yields true state}$$
$$y_{ij} \sim \text{Bernoulli}(z_i \times p_{ij}) \qquad \text{Observation process yields observations}$$

Hence, true occurrence z_i of *G. germanica* at site i is a Bernoulli random variable governed by the parameter ψ (occurrence probability), which is exactly the parameter that most distribution modelers wish they were modeling. The actual gentian observation y_{ij}, detection or not at site i during survey j (or "presence–absence" datum y_{ij}), is another Bernoulli random variable with a success rate that is the product of the actual occurrence of *G. germanica* at that site, z_i, and detection probability p_{ij} at site i during survey j. Hence, at a site where the gentian doesn't occur, $z = 0$, and y must be 0. Conversely, at an occupied site, we have $z = 1$, and *G. germanica* is detected with probability p_{ij}. That is, in the site-occupancy model, the detection probability is expressed *conditional on occurrence*, and the two parameters ψ and p are separately estimable if replicate visits are available.

One way to look at this model in terms of the GLM framework, that features so prominently in this book, is as a hierarchical, coupled logistic regression. One logistic regression describes true occurrence, and the other describes detection, given that the species does occur. Another description of the model is as a nonstandard binomial GLMM with a binary distribution for the random effects—occupied or not occupied—instead of the normal distributions that we assumed in the "standard" binomial GLMM in Chapter 19 as well as most other mixed models in previous chapters.

The above equations describe just the simplest kind of a site-occupancy model; they can readily be extended to more complex cases. First, the ability to model covariate effects is crucial; indeed, covariates can easily be modeled into both the occurrence (ψ) and the detection (p) part of the model in a simple GLM fashion. That is, we can add to the model a statement like this:

$$\text{logit}(\psi_i) = \alpha + \beta^* x_i,$$

where x_i is the value of some occurrence-relevant covariate measured at site i, and α and β are parameters. The same can be done for the observation model also, i.e., the logit transformation of the detection parameter can be modeled by either a site covariate (x_i) or a survey covariate (x_{ij}).

We could also model the effects of many explanatory variables, of polynomial terms or even of splines.

Second, the model, as hitherto described, assumes a so-called closed population, i.e., a site is assumed to be either occupied or not occupied over the entire survey period. For plant surveys conducted during a single season, this may be a sensible assumption. But as soon as surveys are extended over several growing seasons, or the framework is applied to more short-lived or mobile organisms, such as most animals, occupancy status may change within the survey period. Indeed, colonization/extinction dynamics may be a research focus as in metapopulation studies (Hanski, 1998). The solution then is to collect data according to the robust design, where two or more surveys are conducted within a short time period (called a "season"), when the population can be assumed closed (Williams et al., 2002). This is repeated over multiple seasons between which the population may change. For such data, there is a dynamic formulation of the simple site-occupancy model described here to analyze occupancy, colonization, and extinction rates corrected for imperfect detection and to estimate covariate effects on each of these parameters, see MacKenzie et al. (2003, 2006), and Royle and Kéry (2007). In our case, however, we will simulate and analyze data only from a single season and assume a closed population.

20.2 DATA GENERATION

We now simulate the data for our inventory of G. *germanica*. We assume that the 150 sites visited form a sample of a larger number of calcareous grasslands in the Jura. The study objective is to use these data to learn about all calcareous grassland sites in the Jura and to see whether site humidity affects the distribution of G. *germanica*.

```
n.site <- 150          # 150 sites visited
```

We create an arbitrary continuous index for soil humidity, where −1 means dry and 1 means wet, and sort the data for convenient presentation of the results.

```
humidity <- sort(runif(n = n.site, min = −1, max =1))
```

Next, we create the positive true relationship between occurrence probability of G. *germanica* and soil humidity (Fig. 20.2). We do this by a logit-linear regression as customary for binomial responses. We choose the intercept and the slope for this relationship so that about 50% of all sites end up being occupied by the gentian.

```
alpha.occ <- 0          # Logit-linear intercept for humidity on occurrence
beta.occ <- 3           # Logit-linear slope for humidity on occurrence
```

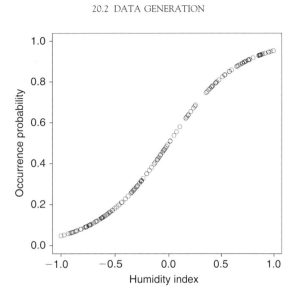

FIGURE 20.2 The unobserved true relationship between humidity and occurrence probability of the Chiltern gentian.

```
occ.prob <- exp(alpha.occ + beta.occ * humidity) / (1 + exp(alpha.occ + beta.occ
* humidity))

plot(humidity, occ.prob, ylim = c(0,1), xlab = "Humidity index", ylab =
"Occurrence probability", main = "", las = 1)

true.presence <- rbinom(n = n.site, size = 1, prob = occ.prob)
true.presence          # Look at the true occupancy state of each site
sum(true.presence)     # Among the 150 visited sites

> true.presence <- rbinom(n = n.site, size = 1, prob = occ.prob)
> true.presence # Look at the true occupancy state of each site
  [1] 0 0 0 1 0 0 0 0 0 0 0 1 0 0 0 0 0 0 1 0 0 0 1 0 0 0 0 0 0 1 0 0 0 0
 [36] 0 0 0 0 0 0 0 0 0 0 0 1 0 1 0 0 1 0 0 0 1 0 0 1 1 0 0 0 0 0 0 1 0
 [71] 0 0 0 1 1 0 0 1 1 1 1 0 1 1 0 1 1 1 1 0 1 0 0 1 0 0 1 1 1 1 1 1 0 1
[106] 1 1 1 0 1 1 1 0 1 1 1 1 1 1 1 1 1 1 1 1 1 1 1 1 1 1 1 1 1 1 1 1 1 1
[141] 0 1 1 1 1 1 1 1 0 1
> sum(true.presence)# Among the 150 visited sites
[1] 75
```

This is the *true state* of the gentian system we are studying, i.e., the realization of the stochastic biological process we're interested in. This state is only imperfectly observable in nature, even for plant populations (Kéry et al., 2006). However, it is what we would like to observe and what we would like to relate to habitat variables such as humidity in our distribution model for *G. germanica*.

Unfortunately, we only observe a degraded image of that true state of nature, where the degradation is because of the fact that *G. germanica* may be overlooked and particularly so at the wetter sites with higher vegetation. So, next we simulate this effect to obtain our actual "presence–absence" observations. After simulating the biological process (which resulted in 150 realizations of true occurrence z_i), we now model the observation process that resides between the true biological process ("truth") and our observations (Fig. 20.3).

```
alpha.p <- 0              # Logit-linear intercept for humidity on detection
beta.p <- -5              # Logit-linear slope for humidity on detection
det.prob <- exp(alpha.p + beta.p * humidity) / (1 + exp(alpha.p + beta.p *
humidity))

plot(humidity, det.prob, ylim = c(0,1), main = "", xlab = "Humidity index", ylab
= "Detection probability", las = 1)
```

Assuming no false-positive errors, i.e., that no other species is erroneously identified as *G. germanica*, the Chiltern gentian can only be detected at sites where it occurs. Hence, the effective detection probability is the product of true occurrence (z_i) and this detection probability (det.prob):

```
eff.det.prob <- true.presence * det.prob
```

Importantly, this effective detection probability or apparent occurrence probability is precisely the quantity modeled by conventional species distribution models (e.g., Elith et al., 2006)! Its expectation is the product of occurrence probability and detection probability, i.e., $\psi \times p$.

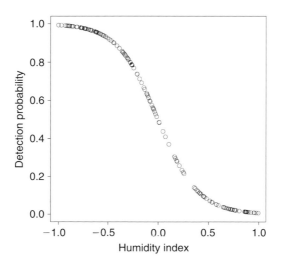

FIGURE 20.3   Relationship between site humidity and detection probability in the Chiltern gentian.

We store the results of each survey, 1 (gentian detected) or 0 (no gentian detected), in an n.site-by-3 matrix and fill it by simulating coin-flips (i.e., drawing Bernoulli trials) with the detection probabilities just computed. We note that this is the first time in the book that we use a two-dimensional array for our data. As a consequence, we will see a double `for` loop in the BUGS code to analyse these data.

```
R <- n.site
T <- 3
y <- array(dim = c(R, T))

# Simulate results of first through last surveys
for(i in 1:T){
    y[,i] <- rbinom(n = n.site, size = 1, prob = eff.det.prob)
}
```

Hence, `y` now contains the results of our simulated surveys to find *G. germanica* at 150 sites.

```
y      # Look at the data
sum(apply(y, 1, sum) > 0) # Apparent distribution among 150 visited sites
> y# Look at the detection/nondetection data
        [,1]  [,2]  [,3]
  [1,]    0     0     0
  [2,]    0     0     0
[ ... ]
[149,]    0     0     0
[150,]    0     0     0
> sum(apply(y, 1, sum) > 0)# Apparent distribution
[1] 31
```

On average (if we simulate this stochastic system many times) our parameter values yield about 42% detected gentian populations. Let's see what a naïve analysis of these observations would tell us about the relationship between humidity and the occurrence of *G. germanica*. (I call this analysis naïve because it omits an important system component, the observation process. The simplest way to analyze this relationship is by a logistic regression of an indicator for "ever detected" (here called `obs`) on humidity.

```
obs <- as.numeric(apply(y, 1, sum) > 0)
naive.analysis <- glm(obs ~ humidity, family = binomial)
summary(naive.analysis)
lin.pred <- naive.analysis$coefficients[1] + naive.analysis$coefficients[2] *
humidity
plot(humidity, exp(lin.pred) / (1 + exp(lin.pred)), ylim = c(0,1), main = "",
xlab = "Humidity index", ylab = "Predicted probability of occurrence", las = 1)
```

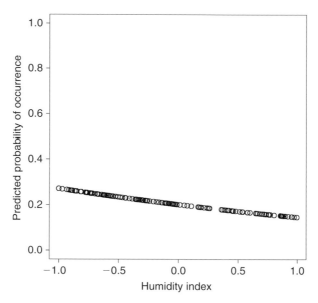

FIGURE 20.4   A naïve analysis of the apparent relationship between occurrence of the Chiltern gentian and humidity. This conventional species distribution model ignores detection probability.

We see that in a naïve analysis, the gentian's preference for wetter sites is totally masked (Fig. 20.4; although in different realizations of the data-generating process, you might also find significant positive or even negative slopes for humidity). Let's see whether a site-occupancy model can do better.

20.3   ANALYSIS USING WinBUGS

A variety of site-occupancy models (e.g., with covariates, for single or multiple seasons, single or multiple species) may be fitted using maximum likelihood in the free Windows-based programs MARK (see *http:// welcome.warnercnr.colostate.edu/~gwhite/mark/mark.htm*) and PRESENCE (see *http://www.mbr-pwrc.usgs.gov/software/doc/presence/presence.html*). R code for obtaining maximum likelihood estimates (MLEs) can be found in the Web appendix of the book by Royle and Dorazio (2008). Furthermore, there is a new R package called unmarked which allows to fit these models using maximum likelihood (Fiske and Chandler, 2010). Hence, here we will directly use WinBUGS to fit the site-occupancy model in a Bayesian mode of inference.

As an added feature, we will perform a posterior predictive check based on the sum of absolute residuals. The resulting Bayesian *p*-value allows us to judge whether the assumed model is appropriate for our data set (based on the selected discrepancy measure).

```
# Define model
sink("model.txt")
cat("
model {

# Priors
 alpha.occ ~ dunif(-10, 10)      # Set A of priors
 beta.occ ~ dunif(-10, 10)
 alpha.p ~ dunif(-10, 10)
 beta.p ~ dunif(-10, 10)
# alpha.occ ~ dnorm(0, 0.01)   # Set B of priors
# beta.occ ~ dnorm(0, 0.01)
# alpha.p ~ dnorm(0, 0.01)
# beta.p ~ dnorm(0, 0.01)

# Likelihood
 for (i in 1:R) {  #start initial loop over the R sites
 # True state model for the partially observed true state
    z[i] ~ dbern(psi[i])              # True occupancy z at site i
    logit(psi[i]) <- alpha.occ + beta.occ * humidity[i]

    for (j in 1:T) {  # start a second loop over the T replicates
       # Observation model for the actual observations
       y[i,j] ~ dbern(eff.p[i,j])   # Detection-nondetection at i and j
       eff.p[i,j] <- z[i] * p[i,j]
       logit(p[i,j]) <- alpha.p + beta.p * humidity[i]

       # Computation of fit statistic (for Bayesian p-value)
       Presi[i,j] <- abs(y[i,j]-p[i,j])   # Absolute residual
       y.new[i,j]~dbern(eff.p[i,j])
       Presi.new[i,j] <- abs(y.new[i,j]-p[i,j])
    }
 }

fit <- sum(Presi[,])# Discrepancy for actual data set
fit.new <- sum(Presi.new[,])   # Discrepancy for replicate data set

# Derived quantities
 occ.fs <- sum(z[])        # Number of occupied sites among 150
}
",fill=TRUE)
sink()

# Bundle data
win.data <- list(y = y, humidity = humidity, R = dim(y)[1], T = dim(y)[2])

# Inits function
zst <- apply(y, 1, max)        #Good starting values for latent states essential !
inits <- function(){list(z = zst, alpha.occ=runif(1, -5, 5), beta.occ = runif(1,
-5, 5), alpha.p = runif(1, -5, 5), beta.p = runif(1, -5, 5))}
```

```
# Parameters to estimate
params <- c("alpha.occ","beta.occ", "alpha.p", "beta.p", "occ.fs", "fit",
"fit.new")

# MCMC settings
nc <- 3
nb <- 2000
ni <- 12000
nt <- 5

# Start Gibbs sampler
out <- bugs(win.data, inits, params, "model.txt", n.chains=nc, n.iter=ni, n.burn
= nb, n.thin=nt, debug = TRUE)
```

Before inspecting the parameter estimates, we first check the adequacy of the model for our data set using a posterior predictive check (Fig. 20.5).

```
plot(out$sims.list$fit, out$sims.list$fit.new, main = "", xlab = "Discrepancy for
actual data set", ylab = "Discrepancy for perfect data sets", las = 1)
abline(0,1, lwd = 2)
```

We then compute a Bayesian *p*-value based on the posterior predictive distributions of our discrepancy measures. In our graph, this corresponds to the proportion of points above the line and for an adequate model, it should be around 0.5 and not approach too much either toward 0 or 1.

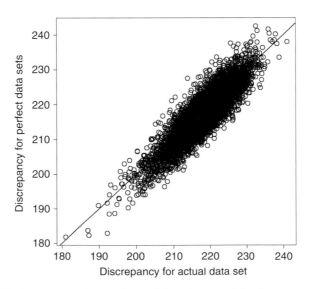

FIGURE 20.5   Posterior predictive check of the adequacy of the site-occupancy model for the gentian data based on the sum of absolute residuals. A well-fitting model has an even number of circles on either side of the 1:1 line.

```
mean(out$sims.list$fit.new > out$sims.list$fit)
> mean(out$sims.list$fit.new > out$sims.list$fit)
[1] 0.3688333
```

The model seems to fit well, so we compare the known true values from the data-generating process with what the site-occupancy analysis has recovered.

```
cat("\n *** Known truth ***\n\n")
alpha.occ ; beta.occ ; alpha.p ; beta.p
sum(true.presence) # True number of occupied sites, to be compared with occ.fs
sum(apply(y, 1, sum) > 0)        # Apparent number of occupied sites
cat("\n *** Our estimate of truth *** \n\n")
print(out, dig = 3)

> cat("\n *** Known truth ***\n\n")

  *** Known truth ***

> alpha.occ ; beta.occ ; alpha.p ; beta.p
[1] 0
[1] 3
[1] 0
[1] -5

> sum(true.presence) # True number of occupied sites, to be compared with occ.fs
[1] 75

> sum(apply(y, 1, sum) > 0) # Apparent number of occupied sites
[1] 31

> cat("\n *** Our estimate of truth *** \n\n")

  *** Our estimate of truth ***

> print(out, dig = 3)
Inference for Bugs model at "model.txt", fit using WinBUGS,
  3 chains, each with 12000 iterations (first 2000 discarded), n.thin = 5
  n.sims = 6000 iterations saved
            mean     sd    2.5%     25%     50%     75%   97.5%  Rhat n.eff
alpha.occ  0.445  0.466  -0.417   0.127   0.427   0.744   1.394 1.001  3900
beta.occ   3.737  1.094   1.769   2.971   3.673   4.444   6.043 1.002  2600
alpha.p   -0.255  0.259  -0.767  -0.432  -0.256  -0.084   0.256 1.001  6000
beta.p    -5.514  0.817  -7.237  -6.030  -5.484  -4.955  -3.988 1.001  6000
occ.fs    76.752  5.812  62.000  74.000  78.000  81.000  85.000 1.001  5800
fit      218.370  6.399 204.297 214.500 218.800 222.600 229.900 1.001  6000
fit.new  217.376  7.246 201.900 212.900 217.700 222.200 230.600 1.001  6000
[ ... ]
```

Thus, the site-occupancy species distribution model succeeds well in recovering the true relationships between humidity and occurrence and detection probability, respectively, which we built into the data. Particularly impressive is its ability to estimate the true number of occupied sites: gentians were only detected at 31 of the known 75 sites where they actually occurred, and the site-occupancy model estimates this number at 76.7, with a 95% credible interval of 62–85. (Note that when analyzing my simulated data set from the book website you will get slightly different results because of Monte Carlo sampling error.)

Finally, we will graphically compare the results from the naïve and the site-occupancy analysis with truth by plotting the true and estimated relationships between occurrence probability of the Chiltern gentian and site humidity (Fig. 20.6). This plot shows the mean predictions under the models; credible intervals could be added if we wished so.

```
plot(humidity, exp(lin.pred) / (1 + exp(lin.pred)), ylim = c(0,1), main = "",
ylab = "Occurrence probability", xlab = "Humidity index", type = "l", lwd = 2,
col = "red", las = 1)
points(humidity, occ.prob, ylim = c(0,1), type = "l", lwd = 2, col = "black")
lin.pred2 <- out$mean$alpha.occ + out$mean$beta.occ * humidity
points(humidity, exp(lin.pred2) / (1 + exp(lin.pred2)), ylim = c(0,1), type =
"l", lwd = 2, col = "blue")
```

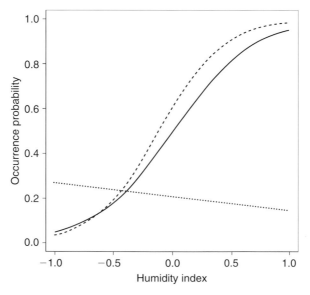

FIGURE 20.6 Comparison of true and estimated relationship between occurrence probability and humidity in the Chiltern gentian (*G. germanica*) under a site-occupancy model (dashed line) and under the naïve approach that ignores detection probability (dotted line). Truth is shown in solid line.

It is evident from Fig. 20.6 that nonaccounting for detection in species distribution models may lead one spectacularly astray. However, some might argue that we have simulated a pathological case, and that one would rarely find such a situation in nature. This may be true. But, we don't know until we have conducted the right analysis. And, we know that even a constant detection probability <1 not only biases the apparent distribution in conventional models but also biases low the strength of covariate relationships. Also, given suitable data, the site-occupancy distribution model can correct for that. This should make it a serious candidate for species distribution modeling when replicate observations are available from at least some sites.

In practice, this very positive conclusion about the model and its implementation in WinBUGS needs to be moderated somewhat. Performance of the model will be inferior with smaller samples (e.g., with fewer sites, a smaller proportion of occupied sites or lower detection probability, or fewer replicate visits) and presumably also in the presence of important unmeasured covariates. On the application side, it must be said that WinBUGS can be painful when fitting these slightly more complex models. For instance, it is essential to provide adequate starting values, in particular, for the latent state (occupancy, the code bit $z = zst$). There may also be prior sensitivity. For instance, altering the set B prior precision from 0.01 to 0.001 may cause WinBUGS to issue one of its dreaded trap messages (e.g., TRAP 66 (postcondition violated)). Sometimes, they are produced even with uniform or normal(0, 0.01) priors. Many other seemingly innocent modeling choices may influence success or failure when fitting a particular model to a given data set.

Hence, for suitable data, site-occupancy models in WinBUGS are great, but a comprehensive analysis of a more complex model may have to be accompanied by a few simulations to check the quality of the inference for the particular case. In addition, sometimes you must be prepared to do quite some amount of painstaking trial and error until the code works. But then, to a large part, this applies quite generally for more complex models, not just site-occupancy models and not just to models fitted using WinBUGS.

20.4 SUMMARY

The site-occupancy model is an extended logistic regression that can estimate true occurrence probability (ψ) and the factors affecting it, while correcting for imperfect detection. The extension is represented by the model component for detection probability (p): conventional logistic regression is a special case, when $p = 1$. Site-occupancy models are the only current framework for inference about species distributions that model true rather than apparent distributions; the latter is the product of occurrence and detection probability (i.e., $\psi \times p$). Our example shows that not accounting for detection probability may lead to spectacularly wrong inferences about the distribution of a species under a conventional, naïve modeling approach. In contrast, the site-occupancy model applied to replicated "presence–absence" data was better able to estimate the true system state (site occupancy and covariate relationship).

EXERCISES

1. *Prior sensitivity*: Conduct a simple prior sensitivity analysis. Compare the inference under the normal and the uniform sets of priors. Plot a histogram of the posterior draws for each of the four primary model parameters to see whether the uniform (−10, 10) priors were not too restrictive to be uninformative.

2. *Site and survey covariates*: We have fitted a site covariate, i.e., one that varied among sites but not among surveys. Incorporate a survey covariate into simulation and analysis code, i.e., one that varies by individual survey. An example might be an inventory conducted by different people with differing and known experience. Experience could be rated on a continuous scale from 0 to 1 and more than one person would be sent to each site. Hint: A sampling covariate has the same 2-D-format as the observed detection/nondetection data.

3. *Swiss hare data*: Collapse the hare counts of one particular year (e.g., 2000) to binary data where a 1 indicates the observation of a "large population" (say, a count ≥10). Estimate the proportion of sites inhabited by a large population, i.e., one capable of producing a count ≥10. A site-occupancy model will correct for the fact that a "large population" may appear small, i. e., produce a count ≥10, or is missed altogether. You may fit a site covariate such as elevation on the occurrence probability of "large" populations (i.e., on ψ).

4. *Simulation study*: Extend the simulation in this chapter to see under which conditions the site-occupancy model is superior in performance to a binomial GLM. Things to vary might be the number of sites (e.g., 20, 50, 150, 500), number of replicate visits (e.g., 2, 3, 5, 10), true average occupancy, and true average detection probability. This is a larger project that could be part of a thesis.

Nonstandard GLMMs 2: Binomial Mixture Model to Model Abundance

21.1 INTRODUCTION

Ecology has been defined as the study of distribution and abundance (Andrewartha and Birch, 1954; Krebs, 2001). However, in nature, neither of them can usually be observed without error, and methods may need to be applied to infer the true states of distribution and abundance from imperfect observations. In Chapter 20, we met a protocol, which we called a *metapopulation design*, where the same quantity was assessed in a similar way across R sites and T temporal replicates. We saw that such a metapopulation design enables the application of site-occupancy models, a kind of nonstandard generalized linear mixed model (GLMM) with binary random effects, to estimate true species distribution free of the distorting effects of detection probability. Temporal replicate observations in a closed system allowed us to resolve the confounding between occurrence and detection.

This chapter features another nonstandard GLMM where the random effects distribution is different from normal, namely Poisson. As for the site-occupancy model, the random effects in this model have a precise biological meaning, which is local population size in this case. Thus, this model estimates abundance corrected for imperfect detection from temporally and spatially replicated *counts*, rather than occurrence from detection/nondetection observations as does the site-occupancy model in Chapter 20.

Our ecological motivation in this chapter is that of Dutch sand lizards (*Lacerta agilis*), one of the most widespread reptiles in large parts of Western Europe (Fig. 21.1). For more than a decade, The Netherlands have had a rather interesting reptile-monitoring program where volunteers walk transects of about 2 km length repeatedly during spring and count all reptiles they see (Kéry et al., 2009). We will generate and analyze data in the format collected in this scheme.

FIGURE 21.1   Pair of sand lizards (*Lacerta agilis*), Switzerland, 2006. (*Photo: T. Ott*)

The typical question in monitoring programs is always "are things getting better or worse?," i.e., is there a trend over time in abundance (or distribution)? The usual type of analysis to answer this question for counts used to be linear regression, but since generalized linear models (GLMs) have become fashionable, some sort of Poisson regression has become the method of choice for the analysis of count data. And, as for other ecological analyses, random effects have become popular and thus, nowadays, inference from many time-series of animal count data is often based on variants of Poisson mixed models (e.g., Link and Sauer, 2002; Ver Hoef and Jansen, 2007).

When repeated counts are available, often only the maximum count is analyzed, although this approach simply throws out valuable information. In this chapter, we make full use of replicated counts and adopt the binomial mixture model (also called the N-mixture model) of Royle (2004a) to estimate true abundance, corrected for imperfect detection (also see Dodd and Dorazio 2004; Royle, 2004b; Kéry et al., 2005; Royle et al., 2005; Dorazio, 2007; Kéry, 2008; Wenger and Freeman, 2008; Joseph et al., 2009). For simplicity, we will consider only data from a single year, and hence, we assume population closure, but the model can also be fitted to multiyear data and a trend in abundance can be estimated directly (Royle and Dorazio, 2008, p. 4–7; Kéry and Royle, 2009; Kéry et al., 2010).

First, the model: we assume that a count y_{ij} at site i and made during survey j comes from a two-stage stochastic process. The first stochastic process is the biological process that distributes the animals among the sites. This process generates the site-specific abundance that we would like to model directly but cannot because we hardly ever see all individuals. The standard statistical model for such data is the Poisson distribution, governed by the intensity (density) parameter λ, which is typically conditional on a few habitat covariates. The result of this first stochastic process is the local, site-specific abundance N_i. Given that true state N_i, the second stochastic process is the observation process which, together with N_i, determines the data actually observed, i.e., the counts y_{ij}. A natural model for the observation process in the presence of imperfect detection is the binomial distribution; given that there are N_i sand lizards present and that each has a probability of p_{ij} to be observed at site i during replicate survey j, the number of lizards actually observed is binomially distributed. Two important consequences are that (1) we typically observe fewer than N_i lizards, and (2) the counts y_{ij} will vary automatically from survey to survey even under identical conditions (Kéry and Schmidt, 2008). Three important assumptions of the binomial mixture model are that of population closure, independent and identical detection probability for all individuals at site i and during survey j, except insofar as differences among sites or surveys are modeled by covariates, and absence of double counts and other false positive errors. The effects of violations of these assumptions are still being investigated (e.g., Joseph et al., 2009).

In summary, the binomial mixture model to estimate abundance from temporally and spatially replicated counts can be written succinctly in just two lines:

$$N_i \sim \text{Poisson } (\lambda) \qquad \text{Biological process yields true state}$$
$$y_{ij} \sim \text{Binomial } (N_i, p_{ij}) \quad \text{Observation process yields observations}$$

It is fascinating to note the similarity of this hierarchical model (Royle and Dorazio, 2006, 2008) for abundance and that for occurrence, the site-occupancy model from Chapter 20. Recognizing that in the site-occupancy model, the observation process may also be described by a binomial distribution (rather than by a Bernoulli), the sole thing that changes when we go from the modeling of occurrence to that of abundance is the distribution used to model the biological process, a Poisson instead of a Bernoulli. The binomial mixture model can be described as a binomial GLMM with a discrete (Poisson-distributed) random effect or alternatively, as a logistic regression for the count observations coupled with a Poisson regression for the imperfectly observed abundances.

Furthermore, as in the site-occupancy model, covariate effects can be modeled into the Poisson parameter λ via a log link function and into the binomial success rate p via the logit link. We can add to the model expressions such as $\log(\lambda_i) = \alpha + \beta * x_i$ and $\text{logit}(p_{ij}) = \alpha + \beta * x_{ij}$, where x_i and x_{ij} are the values of a site-covariate measured at site i (x_i) or of a survey-covariate measured at site i during survey j (x_{ij}). Of course, more than a single covariate could be fitted and the covariates for detection can be of both types.

Before we embark on our usual simulation-analysis exercise, we make two important observations on the abundance parameter of the binomial mixture model. First, even when correcting for imperfect detection (p_{ij}), the interpretation of the abundance parameter N_i is not what we might want it to be: the number of individuals that permanently reside within a well-defined plot of land. The reason for why this is not so is that animals move around, so the effective sampling area will be greater than the nominal sampling area. Hence, the estimate of N_i refers to a larger area, and we don't exactly know the size of it. The magnitude of this discrepancy depends on two things: the typical dispersal distances of the study species and the time frame of the repeated surveys. The discrepancy will be greater for greater dispersal and a longer total survey period. If we want to circumvent this difficulty, other sampling and analysis protocols must be used such as distance sampling (Buckland et al., 2001) or, more recently, spatial capture–recapture methods (Efford, 2004; Borchers and

Efford, 2008; Royle and Dorazio, 2008; Royle and Young, 2008; Efford et al., 2009; Royle et al., 2009).

Second, when animals move through the sampling area randomly, thus in effect violating the closure assumption, it appears that the estimate of N_i does not refer to the number of animals that permanently reside within the sample area, but to the number of animals that ever use an area during the entire sampling period (Joseph et al., 2009). This reasoning is analogous to the reasoning about the interpretation of the occupancy parameter in site-occupancy models in the face of temporary emigration (MacKenzie et al., 2006). In effect, it again makes the effective sampling area larger than the nominal sampling area.

These issues are not a fault of the binomial mixture model; rather, even in the absence of a formal framework for interpreting a count in the light of both the biological *and* the observation process, we never know exactly with which area the count is associated. Nor do we know by how much movement inflates our counts relative to the number of individuals that permanently reside within the area. Thus, a binomial mixture model solves the problem of imperfect detection when interpreting (i.e., analyzing) counts, but the issue of how exactly abundance should be interpreted may still remain a challenge.

21.2 DATA GENERATION

In this example of the analysis of data collected from a metapopulation design, we will choose a different format from that in Chapter 20. Instead of a rectangular or horizontal format (remember the data matrix y_{ij} in the site-occupancy analysis), we will assemble and analyze the data in a vertical format and use a population index covariate to keep track, for each count, of the population it was made in. This format is more convenient when there are many missing values, i.e., in unbalanced designs, where the number of replicate surveys is variable among sites. (It is fairly easy to formulate the site-occupancy model in this format also and the code in the current chapter may be used as a template. For an example of Win-BUGS code for the binomial mixture model in horizontal format, see Kéry, 2008.) Also note the similarity of this to traditional (normal) repeated measures analysis of variance analysis, where some statistics packages allow the replicate observations to be stored in parallel columns (the horizontal format) and others prefer the observations in a single column (the vertical format) with an additional covariate that indexes "subjects." This reiterates the fact that both site-occupancy and binomial mixture models are a kind of repeated-measures analysis.

To simulate our data, we assume that we surveyed 200 sites. We choose a site covariate affecting the abundance of sand lizards and also their detection probability. We take vegetation cover this time and assume that lizard abundance is highest at medium values: too open is bad, but too dense vegetation is also bad. We will model this as a quadratic effect of vegetation density on abundance.

```
n.site <- 200
vege <- sort(runif(n = n.site, min = -1.5, max =1.5))
```

We construct the relationship between vegetation density and abundance (Fig. 21.2).

```
alpha.lam <- 2          # Intercept
beta1.lam <- 2          # Linear effect of vegetation
beta2.lam <- -2         # Quadratic effect of vegetation
lam <- exp(alpha.lam + beta1.lam * vege + beta2.lam * (vege^2))

par(mfrow = c(2,1))
plot(vege, lam, main = "", xlab = "", ylab = "Expected abundance", las = 1)

N <- rpois(n = n.site, lambda = lam)
table(N)                # Distribution of abundances across sites
sum(N > 0) / n.site     # Empirical occupancy

> table(N)# Distribution of abundances across sites
N
```

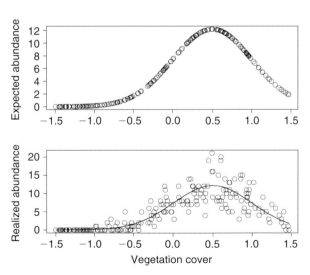

FIGURE 21.2 Expected (top) and realized relationship (bottom) between sand lizard abundance and vegetation cover. The expected abundance is shown as a black line in the bottom panel and is the same as the realized abundance minus Poisson variability.

```
 0  1  2  3  4  5  6  7  8  9  10  11  12  13  14  15  16  19  20  21
57 12 11  7  7 16  8 13 14 12   9   9   7   4   5   3   2   2   1   1

> sum(N > 0) / n.site # Empirical occupancy
[1] 0.715

plot(vege, N, main = "", xlab = "Vegetation cover", ylab = "Realized abundance")
points(vege, lam, type = "l", lwd = 2)
```

This concludes our description of the biological process: we have a random process that distributes the sand lizards across sites, and we assume that the result of this stochastic process at site *i* can be approximated by a conditional Poisson distribution, with rate parameter λ_i, that itself depends on vegetation density x_i in a quadratic fashion.

Next, we need to simulate the observation process, i.e., something like a "machine" that maps abundance N_i onto lizard counts y_{ij}. We assume that the observation process is also affected by vegetation density: denser vegetation reduces the detection probability (Fig. 21.3 upper panel). I note in passing that in the binomial mixture model, detection probability is defined per individual animal, whereas in the occupancy model, it refers to the probability to detect *at least one* among the N_i animals or plants present at a site. In fact, there is a precise mathematical relationship between the two kinds of detection probability, which we do not show here (see Royle and Nichols, 2003; Dorazio, 2007).

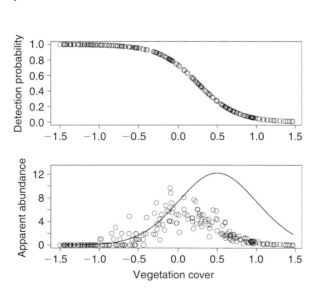

FIGURE 21.3 The relationship between vegetation cover and detection probability (top) and the expected sand lizard counts (=apparent abundance; bottom). Truth is shown as a black line in the bottom panel.

```
par(mfrow = c(2,1))
alpha.p <- 1                # Intercept
beta.p <- -4                # Linear effect of vegetation
det.prob <- exp(alpha.p + beta.p * vege) / (1 + exp(alpha.p + beta.p * vege))
plot(vege, det.prob, ylim = c(0,1), main = "", xlab = "", ylab = "Detection
probability")
```

Now for fun, let's see the expected lizard count at each site in relation to vegetation cover. The expected count at site i is given by the product of abundance N_i and detection probability at that site p_i. And let's put it all together and inspect the truth also.

```
expected.count <- N * det.prob
plot(vege, expected.count, main = "", xlab = "Vegetation cover", ylab = "Apparent
abundance", ylim = c(0, max(N)), las = 1)
points(vege, lam, type = "l", col = "black", lwd = 2) # Truth
```

A conventional analysis would use some sort of Poisson regression and model the expected count or apparent abundance. This is the bell-shaped cloud in the lower panel of Fig. 21.3, where abundance and detection are confounded. Thus, compared with the truth represented by the black line, a conventional analysis will underestimate average abundance and (in our case) estimate maximum abundance at too low of a value of vegetation cover. This is because the Poisson regression does not model abundance but rather the product of expected abundance with detection probability. This is hardly ever made explicit by authors and apparently often not even recognized.

Now let's simulate three replicate counts at each site and look at the data.

```
R <- n.site
T <- 3                       # Number of replicate counts at each site
y <- array(dim = c(R, T))

for(j in 1:T){
    y[,j] <- rbinom(n = n.site, size = N, prob = det.prob)
}
y
```

A species (occurrence) distribution is fundamentally the same as an abundance distribution, but with much reduced information: a species occurs at all sites where abundance $N > 0$ (Royle et al., 2005; Dorazio, 2007). Hence, any model of abundance is also a model of species distribution. It is seldom useful to think of distribution as something separate from abundance.

```
sum(apply(y, 1, sum) > 0)        # Apparent distribution (proportion occupied sites)
sum(N > 0)                       # True occupancy

> sum(apply(y, 1, sum) > 0)
[1] 126
> sum(N > 0)
[1] 143
```

Now stack the replicated counts on top of each other for a vertical data format (i.e., convert the matrix to a vector)

```
C <- c(y)
```

We also need a site index and a vegetation covariate that have the same length as the variable C (for the observation model, i.e., to model *p*; see WinBUGS code in section 21.3.). We will denote them by a p suffix in the variable name.

```
site = 1:R                    # 'Short' version of site covariate
site.p <- rep(site, T)        # 'Long' version of site covariate
vege.p <- rep(vege, T)        # 'Long' version of vegetation covariate
cbind(C, site.p, vege.p)      # Check that all went right
```

Here is a quick and dirty conventional analysis of the maximum counts assuming a Poisson distribution for the max(count) (A slightly better alternative would, perhaps, be to assume a normal distribution for the mean of the counts.):

```
max.count <- apply(y, 1, max)
naive.analysis <- glm(max.count ~ vege + I(vege^2), family = poisson)
summary(naive.analysis)
lin.pred <- naive.analysis$coefficients[1] + naive.analysis$coefficients[2] *
vege + naive.analysis$coefficients[3] * (vege*vege)
```

We compare truth and the naïve analysis in a graph (Fig. 21.4):

```
par(mfrow = c(1,1))
plot(vege, max.count, main = "", xlab = "Vegetation cover", ylab = "Abundance or
count", ylim = c(0,max(N)), las = 1)
points(vege, lam, type = "l", col = "black", lwd = 2)
points(vege, exp(lin.pred), type = "l", col = "red", lwd = 2)
```

Clearly, the predictions under the naïve analysis yield a biased picture of the relationship between abundance and vegetation cover because sand lizards are easier to see in more open vegetation.

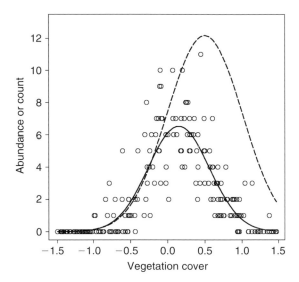

FIGURE 21.4   Relationship between abundance of Dutch sand lizards and vegetation cover as inferred by a naïve analysis not accounting for detection probability. Circles show the maximum count at each site and the black solid line the predicted relationship under the naïve model. Truth is a dashed line.

21.3 ANALYSIS USING WinBUGS

Now let's see how the binomial mixture model can do better. As for the site-occupancy model, a variety of binomial mixture models can be fitted using maximum likelihood in the free Windows programs MARK and PRESENCE. R code for obtaining maximum likelihood estimates under the model can be found in Kéry et al. (2005), Royle and Dorazio (2008) and its Web appendix, and Wenger and Freeman (2008). Furthermore, the model can be fitted using functions in the new R package unmarked (Fiske and Chandler, 2010). We show a Bayesian solution here.

```
# Define model
sink("BinMix.txt")
cat("
model {

# Priors
  alpha.lam ~ dnorm(0, 0.1)
  beta1.lam ~ dnorm(0, 0.1)
  beta2.lam ~ dnorm(0, 0.1)
  alpha.p ~ dnorm(0, 0.1)
  beta.p ~ dnorm(0, 0.1)
```

```
# Likelihood
# Biological model for true abundance
  for (i in 1:R) {              # Loop over R sites
    N[i] ~ dpois(lambda[i])
    log(lambda[i]) <- alpha.lam + beta1.lam * vege[i] + beta2.lam * vege2[i]
  }

# Observation model for replicated counts
  for (i in 1:n) {              # Loop over all n observations
      C[i] ~ dbin(p[i], N[site.p[i]])
      logit(p[i]) <- alpha.p + beta.p * vege.p[i]
  }

# Derived quantities
  totalN <- sum(N[])          # Estimate total population size across all sites
}
",fill=TRUE)
sink()

# Bundle data
R = dim(y)[1]
n = dim(y)[1] * dim(y)[2]
vege2 = (vege * vege)
win.data <- list(R = R, vege = vege, vege2 = vege2, n = n, C = C, site.p =
site.p, vege.p = vege.p)
```

As for the site-occupancy model, clever starting values for the latent states (the N_i's) are essential. We use the maximum count at each site as a first guess of what N might be and add 1 to avoid zeros. WinBUGS cannot use zeroes for the N value for the binomial and will crash if you initialize the model with zeroes.

```
# Inits function
Nst <- apply(y, 1, max) + 1
inits <- function(){list(N = Nst, alpha.lam=rnorm(1, 0, 1), beta1.lam=rnorm(1, 0,
1), beta2.lam=rnorm(1, 0, 1), alpha.p=rnorm(1, 0, 1), beta.p=rnorm(1, 0, 1))}

# Parameters to estimate
params <- c("N", "totalN", "alpha.lam", "beta1.lam", "beta2.lam", "alpha.p",
"beta.p")

# MCMC settings
nc <- 3
nb <- 200
ni <- 1200
nt <- 5
```

```
# Start Gibbs sampler
out <- bugs(win.data, inits, params, "BinMix.txt", n.chains=nc, n.iter=ni, n.burn
= nb, n.thin=nt, debug = TRUE)
```

Here is a first note on the practical implementation of such slightly more complex models in WinBUGS: This code works fine and appears to converge surprisingly quickly for our data set. But as an illustration of how "difficult" WinBUGS can sometimes be, try widening the range of the priors by increasing some or all of the precisions from 0.1 to 0.01: WinBUGS will crash. Many modeling choices that are not wrong, but simply not chosen in an optimal manner, can throw you off the track in your attempts to exploit the great modeling freedom that WinBUGS gives you in principle. It is true that with experience, the reasons for many crashes can be diagnosed, but for a beginner, they may represent major stumbling blocks.

Let's now compare the known true values with what the analysis has recovered.

```
cat("\n *** Our estimate of truth *** \n\n")
print(out, dig = 2)

cat("\n *** Compare with known truth ***\n\n")
alpha.lam    ;    beta1.lam    ;    beta2.lam    ;    alpha.p    ;    beta.p
sum(N)            # True total population size across all sites
sum(apply(y, 1, max))        # Sum of site max counts

> cat("\n *** Our estimate of truth *** \n\n")

 *** Our estimate of truth ***

> print(out, dig = 2)
Inference for Bugs model at "BinMix.txt", fit using WinBUGS,
 3 chains, each with 1200 iterations (first 200 discarded), n.thin = 5
 n.sims = 600 iterations saved
```

	mean	sd	2.5%	25%	50%	75%	97.5%	Rha	n.eff
[...]									
totalN	957.01	160.99	683.95	836.00	948.00	1066.00	1282.22	1.01	310
alpha.lam	2.47	0.18	2.11	2.35	2.48	2.59	2.80	1.00	340
beta1.lam	0.71	0.23	0.27	0.54	0.72	0.85	1.17	1.01	480
beta2.lam	−2.81	0.23	−3.25	−2.97	−2.80	−2.65	−2.33	1.02	150
alpha.p	0.14	3.16	−5.80	−2.10	−0.06	2.57	6.32	1.00	600
beta.p	−0.14	2.95	−5.95	−2.15	−0.17	1.96	5.76	1.00	600
deviance	5844.93	4799.46	1385.70	2547.73	4475.00	7278.66	18910.75	1.00	600

```
[ ... ]
>
> cat("\n *** Compare with known truth ***\n\n")
```

```
*** Compare with known truth ***
> alpha.lam   ;   beta1.lam   ;   beta2.lam   ;   alpha.p   ;   beta.p
[1] 2
[1] 2
[1] -2
[1] 1
[1] -4
> sum(N) # True total population size across all sites
[1] 1073
> sum(apply(y, 1, max))# Sum of site max counts
[1] 481
```

With truth being 1073 sand lizards, the estimate of total N (957 lizards) *appears* decent with respect to the sum of the max counts across all 200 sites, which was only 481. However, there is a slight correspondence for the coefficients in the biological process model (alpha.lam, beta1.lam, beta2.lam), but no similarity at all for the coefficients in the observation process model … That is disappointing! What has happened? Note again that WinBUGS claims that the chains have converged.

Seeing these results for the first time, I couldn't believe that this would go wrong because I knew that these parameters should all be identifiable (and were so for an only slightly different model in Kéry, 2008). I tried out several things to see whether I could get a better result; I increased sample sizes (e.g., to R = 2000 and T = 10), dropped the quadratic term in the biological process model, but none of those helped. In the end, I tried out an alternative set of uniform priors instead of the fairly uninformative normals used previously. I also avoided the WinBUGS logit and defined that function myself (see WinBUGS tricks in the Web appendix). And this worked!

That is, I fitted the following version of the model, where I also added code to assess model goodness of fit using a posterior predictive check for a Chi-square discrepancy measure. Convergence in a binomial mixture model is attained notoriously much more slowly than, say, in a site-occupancy model. Therefore, I ran considerably longer chains.

```
# Define model with new uniform priors
sink("BinMix.txt")
cat("
model {

# Priors (new)
  alpha.lam ~ dunif(-10, 10)
  beta1.lam ~ dunif(-10, 10)
  beta2.lam ~ dunif(-10, 10)
  alpha.p ~ dunif(-10, 10)
  beta.p ~ dunif(-10, 10)
```

```
# Likelihood
# Biological model for true abundance
 for (i in 1:R) {
   N[i] ~ dpois(lambda[i])
   log(lambda[i]) <- alpha.lam + beta1.lam * vege[i] + beta2.lam * vege2[i]
 }

# Observation model for replicated counts
 for (i in 1:n) {
   C[i] ~ dbin(p[i], N[site.p[i]])
   lp[i] <- alpha.p + beta.p * vege.p[i]
   p[i] <- exp(lp[i])/(1+exp(lp[i]))
 }

# Derived quantities
 totalN <- sum(N[])

# Assess model fit using Chisquare discrepancy
 for (i in 1:n) {

# Compute fit statistic for observed data
     eval[i]<-p[i]*N[site.p[i]]
     E[i] <- pow((C[i] - eval[i]),2) / (eval[i] + 0.5)

# Generate replicate data and compute fit stats for them
     C.new[i] ~ dbin(p[i], N[site.p[i]])
     E.new[i] <- pow((C.new[i] - eval[i]),2) / (eval[i] + 0.5)
 }
 fit <- sum(E[])
 fit.new <- sum(E.new[])
}
",fill=TRUE)
sink()

# Parameters to estimate
params <- c("N", "totalN", "alpha.lam", "beta1.lam", "beta2.lam", "alpha.p",
"beta.p", "fit", "fit.new")

# MCMC settings
nc <- 3
nb <- 10000
ni <- 60000
nt <- 50                     # Takes about 20 mins on my laptop

# Start Gibbs sampler
out <- bugs(win.data, inits, params, "BinMix.txt", n.chains=nc, n.iter=ni, n.burn
= nb, n.thin=nt, debug = TRUE)
```

Because convergence in Bayesian analyses of binomial mixture models may be hard to achieve, we first check whether this run has converged. We find that the specifications of this Markov chain Monte Carlo (MCMC) run seem to have been long enough. The Markov chains of all primary, structural model parameters seem to have converged, only those for some latent (local abundance N_i) parameters have not converged. Their convergence is of lesser concern, unless of course the chains for local abundance of your favorite population happened not to converge.

```
print(out, dig = 3)
which(out$summary[,8] > 1.1)
```

Next, we check whether the uniform prior on beta.p was not too restrictive by producing a histogram of the posterior (the posterior of this parameter seemed to come closest to the bounds of the uniform prior). However, there is no indication for any mass being piled up toward one of the bounds (Fig. 21.5), and therefore, we conclude that there was no undue influence of this prior on our inference.

```
hist(out$sims.list$beta.p, col = "grey", main = "", xlab = "")
```

Next, we check the adequacy of the model for the data set first using a posterior predictive check, before inspecting the parameter estimates (Fig. 21.6):

```
plot(out$sims.list$fit, out$sims.list$fit.new, main = "", xlab = "Discrepancy
measure for actual data set", ylab = "Discrepancy measure for perfect data sets")
abline(0,1, lwd = 2, col = "black")
```

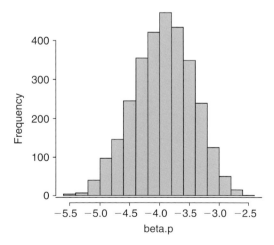

FIGURE 21.5 Posterior distribution of the slope estimate of sand lizard detection probability on vegetation cover (beta.p).

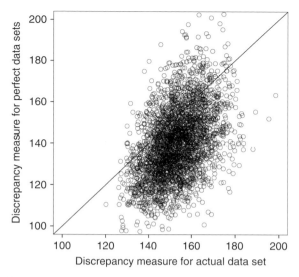

FIGURE 21.6   Posterior predictive check of the binomial mixture model using a Chi-square discrepancy.

```
mean(out$sims.list$fit.new > out$sims.list$fit)
> mean(out$sims.list$fit.new > out$sims.list$fit)
[1] 0.1906667
```

Both the graphical check and the Bayesian *p*-value, indicate an adequate model for our data set, so we inspect the parameter estimates and compare them with truth in the data-generating process.

```
cat("\n *** Our estimate of truth *** \n\n")
print(out, dig = 2)

cat("\n *** Compare with known truth ***\n\n")
alpha.lam    ;    beta1.lam    ;    beta2.lam    ;    alpha.p    ;    beta.p
sum(N)              # True total population size across all sites
sum(apply(y, 1, max))        # Sum of site max counts

> cat("\n *** Our estimate of truth *** \n\n")

 *** Our estimate of truth ***

> print(out, dig = 2)
Inference for Bugs model at "BinMix.txt", fit using WinBUGS,
 3 chains, each with 60000 iterations (first 10000 discarded), n.thin = 50
 n.sims = 3000 iterations saved
            mean     sd    2.5%     25%     50%     75%    97.5% Rhat  n.eff
N[1]        0.00   0.00    0.00    0.00    0.00    0.00    0.00 1.00      1
 [ ... ]
```

```
N[200]          2.26    3.07    0.00    0.00    1.00    3.00    11.00 1.03    290
totalN       1014.12  268.27  683.00  822.00  942.00 1136.25 1741.17 1.02    160
alpha.lam       1.96    0.07    1.82    1.91    1.96    2.00     2.10 1.00   1300
beta1.lam       1.81    0.26    1.34    1.62    1.79    1.99     2.36 1.01    190
beta2.lam      -1.99    0.33   -2.64   -2.22   -1.99   -1.76    -1.35 1.01    280
alpha.p         1.12    0.17    0.78    1.02    1.13    1.24     1.44 1.01    310
beta.p         -3.94    0.49   -4.94   -4.28   -3.92   -3.59    -3.02 1.01    260
fit           153.07   10.64  133.10  145.70  152.70  160.10   175.10 1.00    550
fit.new       139.98   16.73  110.60  128.00  139.10  150.60   176.20 1.01    290
[ ... ]
```

```
DIC info (using the rule, pD = var(deviance)/2)
pD = 209.2 and DIC = 1161.9
DIC is an estimate of expected predictive error (lower deviance is better).
>
> cat("\n *** Compare with known truth ***\n\n")

 *** Compare with known truth ***

> alpha.lam    :   beta1.lam    :    beta2.lam    :   alpha.p    :    beta.p
[1] 2
[1] 2
[1] -2
[1] 1
[1] -4

> sum(N) # True total population size across all sites
[1] 1073

> sum(apply(y, 1, max))# Sum of site max counts
[1] 481
```

Our model recovered adequate parameter estimates for the covariate relationships and estimated a total population size across all 200 sites of 1014 sand lizards (95% CI: 683–1741). As a comparison, truth was 1073 lizards, and the sum of the max counts, a conventional estimate of total population size, was only 481.

Figure 21.7 gives a graphical comparison between the parameter estimates under the binomial mixture model and the values of the associated data-generating parameters. It shows again that the model does a good job at estimating abundance.

```
par(mfrow = c(3,2))
hist(out$sims.list$alpha.lam, col = "grey", main = "alpha.lam", xlab = "")
abline(v = alpha.lam, lwd = 3, col = "black")
hist(out$sims.list$beta1.lam, col = "grey", main = "beta1.lam", xlab = "")
```

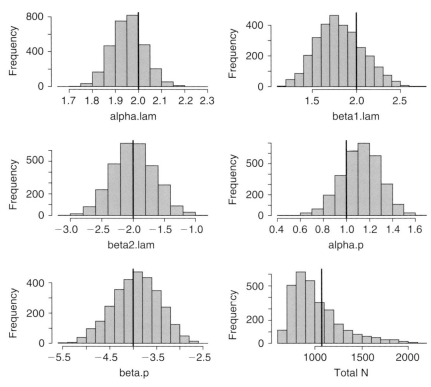

FIGURE 21.7 Comparison of estimates under the binomial mixture model (posterior distributions) and truth in the data-generating algorithm (black line) for six estimands.

```
abline(v = beta1.lam, lwd = 3, col = "black")
hist(out$sims.list$beta2.lam, col = "grey", main = "beta2.lam", xlab = "")
abline(v = beta2.lam, lwd = 3, col = "black")
hist(out$sims.list$alpha.p, col = "grey", main = "alpha.p", xlab = "")
abline(v = alpha.p, lwd = 3, col = "black")
hist(out$sims.list$beta.p, col = "grey", main = "beta.p", xlab = "")
abline(v = beta.p, lwd = 3, col = "black")
hist(out$sims.list$totalN, col = "grey", , main = "Total N", xlab = "")
abline(v = sum(N), lwd = 3, col = "black")
```

Here is a second note on the practical implementation of such slightly more complex models in WinBUGS: We saw that a relatively slight change (here, in the priors) had a very large effect on the inference. This could also be called an example of prior sensitivity of the inference. It is definitely a good idea to check the Bayesian analysis of more complex models by exploring "neighboring model regions." Vary the likelihood (e.g., the covariates that are in or not), priors, model parameterization or other things slightly and see whether your inference is robust.

In our case, we now get rather decent estimates fairly close to the known truth. Hence, we illustrate a few further inferences that can be made under the model by looking at a few further posterior distributions. In Fig. 21.7, we have seen those for some primary parameters of the model. One of the most interesting things in the binomial mixture model is that site-specific estimates, N_i, can be obtained. Let's now have a look at these estimates of local abundance for a random sample of sites (Fig. 21.8).

```
sel <- sort(sample(1:200, size = 4))
sel

par(mfrow = c(2,2))
hist(out$sims.list$N[,sel[1]], col = "grey", xlim = c(Nst[sel[1]]-1,
max(out$sims.list$N[,sel[1]])), main = "Site 48", xlab = "")
abline(v = Nst[sel[1]]-1, lwd = 3, col = "red")
```

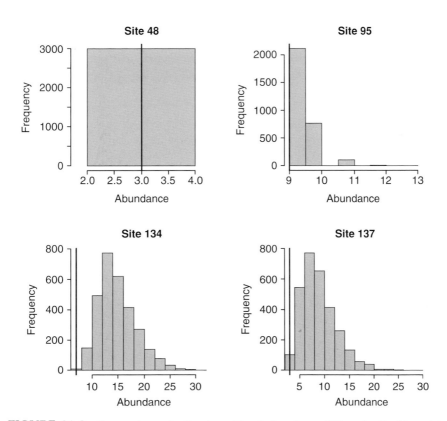

FIGURE 21.8 Comparison of estimates of local abundance (N_i) under the binomial mixture model (posterior distributions) and the maximum count (black line) at a sample of four sites.

```
hist(out$sims.list$N[,sel[2]], col = "grey", xlim = c(Nst[sel[2]]-1,
max(out$sims.list$N[,sel[2]])), main = "Site 95", xlab = "")
abline(v = Nst[sel[2]]-1, lwd = 3, col = "red")

hist(out$sims.list$N[,sel[3]], col = "grey", xlim = c(Nst[sel[3]]-1,
max(out$sims.list$N[,sel[3]])), main = "Site 134", xlab = "")
abline(v = Nst[sel[3]]-1, lwd = 3, col = "red")

hist(out$sims.list$N[,sel[4]], col = "grey", xlim = c(Nst[sel[4]]-1,
max(out$sims.list$N[,sel[4]])) , main = "Site 137", xlab = "")
abline(v = Nst[sel[4]]-1, lwd = 3, col = "red")

> sel
[1]     48    95    134    137
```

The posterior distributions show the likely size of the local populations (N_i) of sand lizards at sites number 48, 95, 134, and 137. We can compare this to the observed data at these sites:

```
y[sel,]
> y[sel,]
      [,1] [,2] [,3]
[1,]   3    3    3
[2,]   6    7    9
[3,]   7    7    3
[4,]   1    3    2
```

And since we know truth, why not have a look at it? Here are the true population sizes at these sites:

```
N[sel]
> N[sel]
[1] 3 9 21 8
```

Compare this with the estimates of these local N_i:

```
print(out$mean$N[sel], dig = 3)
> print(out$mean$N[sel], dig = 3)
[1]  3.00  9.34  15.31  9.31
```

Finally, Fig. 21.9 shows a comparison of the relationship between sand lizard abundance and vegetation cover using a naïve analysis and under the binomial mixture model.

```
par(mfrow = c(1,1))
plot(vege, N, main = "", xlab = "Vegetation cover", ylab = "Abundance", las = 1,
ylim = c(0,max(N)))
```

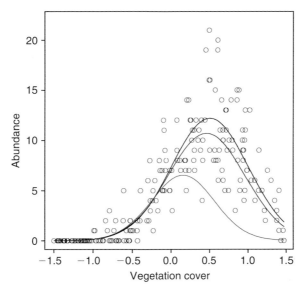

FIGURE 21.9 Comparison of the estimated abundance–vegetation relationship in Dutch sand lizards under a naïve approach that ignores imperfect detection (red) with that under the binomial mixture model (blue). Truth is shown in black: circles are the realized abundances at each site (N_i) and the black line is their expectation (as in Fig. 21.1).

```
points(sort(vege), lam[order(vege)], type = "l", col = "black", lwd = 2)
points(vege, exp(lin.pred), type = "l", col = "red", lwd = 2)
BinMix.pred <- exp(out$mean$alpha.lam + out$mean$beta1.lam * vege +
out$mean$beta2.lam * (vege*vege))
points(vege, BinMix.pred, type = "l", col = "blue", lwd = 2)
```

21.4 SUMMARY

Binomial mixture modeling in metapopulation designs offers great opportunities for the estimation of animal or plant abundance corrected for imperfect detection probability (p). Essentially, the model is simply a generalized version of the familiar Poisson regression model that accommodates imperfect detection; when $p = 1$, we are back to a classical Poisson generalized linear model. However, it is more complex than a simpler Poisson regression and even more so than a conventional hierarchical Poisson regression or GLMM (see Chapter 16). There are many ways in which the practical implementation in WinBUGS may fail, and a lot of trial and error and model checking may be required. Nevertheless, if replicate count data are available from a number of sites, you should absolutely try out this new and exciting model.

EXERCISES

1. *Survey covariates*: In metapopulation designs, we frequently have detection-relevant covariates that vary by site *and* survey (i.e., survey covariates). For our Dutch sand lizards, one such covariate is ambient temperature (Kéry et al., 2009): presumably, lizard activity depends on the temperature and this affects their detection probability. Modify the data generation code as well as the WinBUGS model to include the effects of a temperature covariate.

2. *Prior sensitivity*: Play around with prior settings in the last model we ran. Change the uniform distributions to have a very wide range and see whether the model converges. Conversely, set the range very narrow and see whether the inference is affected, i.e., whether the parameter estimates are changed.

3. *Swiss hare data*: Fit the binomial mixture model to the hare data from a single year (e.g., 2000) and see whether there is a difference in the probability of detection in grassland and arable areas. By what proportion will mean or max counts underestimate true population size?

4. *Simulation exercise*: The binomial mixture model was described in 2004 (Royle, 2004a) and so is still fairly young. Five years later, it had been applied in hardly more than 10 publications, and much remains to be found out about the model that can be tackled by simulation studies. For instance, doing a simulation study (vary R, T, covariate effects, and other things) for models with covariates similar as what was suggested for the site-occupancy model (Exercise 4 in Chapter 20) might be worthwhile. This is another serious study that could easily yield a decent article or thesis chapter.

Conclusions

Now that you are through, it is perhaps worth looking back and seeing where you have come. I hope that you have achieved four things. First, you have achieved some practical understanding for how Bayesian inference with vague priors works and why Bayesian inference is simply so *useful* (Link and Barker, 2010). Second, you have gained plenty of practice with WinBUGS, the most widely used general-purpose Bayesian software. Third, I imagine that many of you have obtained a much deeper understanding for what you have been doing for years: fitting linear, generalized linear, and mixed models. In addition to the insights provided by the simulation of data sets, it is the natural way of model specification in the BUGS language that makes WinBUGS uniquely suitable to really understand these models. So, funny perhaps for a book about applied Bayesian modeling, I think that one of its greatest benefits for you may be something more general: an improved understanding for linear models and their extensions.

Finally, and fourth, I hope you have gained a taste for modern statistical modeling. In statistics classes at university, many ecologists have only seen a sad caricature of statistics. We were taught to think in terms of a decision tree for black-box procedures. The tree started with a question like "Are the data normally distributed?" and its terminal branches prescribed a t-test or a Kruskal–Wallis test or an analysis of covariance with homogeneous slopes (or else we were in deep trouble …). And then, a p-value popped up somewhere, and if it was <0.05, life was good. This is a terrible way of doing statistics, boring and devoid of any creative energy. It also does great injustice to the inspiring activity of trying to make sense of incomplete and imperfect observations made in a noisy world. Perhaps it should come as no surprise that many "sensible" people hate and distrust statistics.

To me, collaborating with some statistician colleagues, and especially starting to use WinBUGS, has been an eye-opener. I have come to see data analysis as equivalent with the creation of a statistical model that attempts to emulate the main features of the stochastic process that could conceivably have generated our data. We then study features of our statistical model representation of that part of the world we are interested in, such as body mass of peregrines, snout-vent length of snakes, population size of lizards, or the distribution of a plant. By doing so, we hope to learn something about that world, something that may be hidden underneath the tangle of detail, distortion, and noise that is a hallmark of most data sets.

Many modern statisticians build their models in an organic way, without ever thinking about that old decision tree that many of us ecologists have come to learn by heart. They choose their ingredients from a vast array and assemble their models in a modular way: take a little of this distribution and add a little of that feature to the linear predictor. This is a much more creative and inspiring act than simply working our way along the branches of that old tree. And one that likely leads to much improved inference and mechanistic insight, since the model can be adapted to the particular situation at hand rather than the converse, with us having to shoehorn reality into the conceptual box provided by our statistics program. Thus, modern statistical analysis means to build one's own custom models.

Unfortunately, I doubt whether most ecologists will ever be sufficiently numerate to fit their custom models by maximum likelihood, at least, if they have to specify themselves the likelihood of their model in an explicit way. This is where WinBUGS comes in. It is the only software I know that allows the average, somewhat numerate ecologist to conduct his or her own creative statistical modeling. Therefore, again, I believe that WinBUGS has the potential to free the statistical modeler in us. Even without the benefits of the Bayesian paradigm of statistical inference, this would be enough for me to recommend WinBUGS to any ambitious quantitative ecologist.

Where will you go from here? I hope that my book enables you to better tackle more advanced books on ecological modeling, such as by Royle and Dorazio (2008), King et al. (2009), or Link and Barker (2010). Much of this advanced modeling is based on the powerful notion of *hierarchical models* (HMs; e.g., Royle and Dorazio, 2006, 2008). HMs describe observed data as the result of a series of linked stochastic processes, whose outcomes may be observed or unobserved, i.e., latent. HMs are very powerful to understand and predict complex ecological systems. They can be fitted also using frequentist methods (e.g., Lee et al., 2006; Ponciano et al., 2009), but their implementation in WinBUGS is much more straightforward and arguably easier to understand for an ecologist. The reason for this is that in the BUGS language, a complex model is naturally broken apart into hierarchically linked submodels. Indeed, in this book, we have seen many instances of HMs: all the models containing random effects and more particularly the site-occupancy and binomial mixture models for inference about metapopulation distribution and abundance (Chapters 20 and 21). In these final chapters, we have just about reached the level of modeling where the book by Royle and Dorazio (2008) starts.

Another direction that you may want to explore is nonlinear models and especially nonlinear mixed models (Pinheiro and Bates, 2000). In this book, we have exclusively adopted linear statistical models, but there is no reason to restrict the deterministic part of our system description to be additive. Nonlinear models may be more realistic representations of a study system and may yield better predictions, especially outside of the observed range of covariates.

Finally, I have no doubts that many Bayesians would accuse me of being Bayesian only in a purely opportunistic way. In this book, we have used the Bayesian computing machinery as a handy way of fitting sometimes complex models without ever really taking advantage of that true hallmark of Bayesian inference: the ability to formally combine the information contained in the data with all available knowledge about the study system. That is, we don't really use informative priors. This is true, and I believe that much can be gained by not feigning ignorance about the system analyzed in the way that inference by maximum likelihood or Bayesian analysis with vague priors does. Indeed, especially with the small data sets that are so typical for ecology, much may be gained in terms of precision of the estimates and parameter estimability when informative priors are used. It seems likely that in the future, we will increasingly see Bayesian analyses with informative priors.

Perhaps WinBUGS may not allow you to go all the way there. WinBUGS may be too slow for your data set, you may never get convergence, be caught in a trap, or just lose the patience when WinBUGS behaves just like a 15-year old. Also, new programs are likely to be developed and improved in the future, such as OpenBugs (*http://mathstat.helsinki.fi/openbugs/*), AD Model Builder (*http://admb-foundation.org/*), JAGS (*http://www-fis.iarc.fr/~martyn/software/jags/*), PyMC (*http://code.google.com/p/pymc/*), among others, and they may eventually be even more accessible to ecologists than WinBUGS is now and offer greater computing power. Or, you may learn how to code your own Markov chain Monte Carlo samplers and perhaps use a general programming language such as Fortran or C++, which may speed up your Bayesian computations by orders of magnitude relative to what you can achieve in WinBUGS. But even then, I believe that for many of you, WinBUGS will have achieved one important thing: it will have changed forever the way in which you think about, and conduct, statistical modeling.

A List of WinBUGS Tricks

I provide here a list of tips that hopefully allow you to love WinBUGS more unconditionally. I would suggest you skim over the list now and then refer back to it later as necessary.

1. *Do read the manual*: WinBUGS may not have the best documentation available for a software, but its manual is nevertheless very useful. Be sure to at least skim over most of it once when you start getting into WinBUGS (i.e., now for many of you), so you may remember that the manual has something to say about a particular topic when you need it. Don't forget the sections entitled "Tricks: Advanced Use of the BUGS Language" and "Tips and Troubleshooting."

2. *How to begin*: When starting a new analysis, always start from a template of a similar analysis. Only ever try to write an analysis from scratch if you want to test yourself.

3. *Initial values 1*: The wise choice of inits can be the key to success or failure of an analysis in WinBUGS, although we don't see this so much in the fairly simple models in this book (the nonstandard generalized linear mixed models in Chapters 20 and 21 are an exception). With more complex models, WinBUGS needs to start the Markov chains not too far away from their stationary distribution or it will crash or not even start to update. Of course, the requirement to start the chains close to the solution goes counter the requirement to start them at dispersed places to assess convergence, so some reasonable intermediate choice is important.

4. *Initial values 2*: Inits must not be outside of the possible range of a parameter. For instance, negative inits for a parameter that has prior mass only for positive values (such as a variance) will cause WinBUGS to crash and so do inits outside of the range of a uniform prior.

5. *Be ignorant, but not too ignorant*: When you want your Bayesian inference to be dominated by your data and choose priors intended to be vague, don't specify too much ignorance, otherwise traps may result or convergence may not be achieved. For instance, don't specify the limits of a uniform prior or the variance of a normal prior to be too wide.

6. *Missing values (NAs)*: NAs are not an issue in this book, since they do not occur in our "perfect" data sets. However, in your real-world data sets, there will always be NAs. In WinBUGS, NAs are dealt with less automatically than in conventional stats programs with which you are likely familiar; hence, it is important to know how to deal with them: briefly, missing responses (i.e., missing y s) are no problem, but NAs in the explanatory variables (the x s) need attention. A missing response is simply estimated, and indeed, adding missing responses for selected covariate values is one of the simplest ways to form predictions for desired values of explanatory variables. On the other hand, a missing explanatory variable must either be replaced with some number, e.g., the mean observed value for that variable, or else given some prior distribution. In general, the former is easier and should not pose a problem unless the number of missing x s is large.

7. *Think in a box (and know your box)*: When coding an analysis in WinBUGS, you often have to deal with data that come in arrays that have more than one dimension. For instance, when analysing animal counts from different sites, over several years, and taken at various months in a year, it may be useful to format them into a three-dimensional array. Some covariates of such an analysis will then have two or even three dimensions, too. Obviously, you must then be absolutely clear about the dimensions of theses "boxes" in which your data are and not get confused by the indexing of the data. In my experience, knowing how to format data into such arrays and then not getting lost is one of the most difficult things to learn about the routine use of WinBUGS.

8. *Know your box (2)*: In "serious" analyses, your modeling often requires the data to be formatted in some multidimensional array. For instance, for a multispecies version of a site-occupancy model (see Chapter 20), you have at least three dimensions corresponding to species (i), site (j), and replicate survey (k). It appears that how you build your array and, especially, how you loop over that array in the definition of the likelihood can make a huge difference in terms of the speed with which your Markov chains in WinBUGS evolve. It appears that you should loop over the longest dimension first and over the shortest last. For instance, if you have data from 450 sites, 100 species, and for two surveys each, then it appears best to format the data as $y[j, i, k]$ and then loop over sites (j) first, then over species (i), and finally over replicate surveys (k).

9. *Don't define things twice*: Every parameter in WinBUGS can only be defined once. For instance, writing `y ~ dnorm(mu, tau)` and then adding `y[3] <- 5` will cause an error. There is a single exception to this rule, and that is the transformation of the response by some

function such as the `log()` or `abs()`. So to conduct an analysis of a log-transformed response, you may write `y <- log(y)` and then `y ~ dnorm(mu, tau)`. Beware of inadvertently defining quantities multiple times when erroneously putting them within a loop that they don't belong.

10. *logit function*: In more complex models, I have fairly often experienced problems when using WinBUGS' own logit function, for instance, with achieving convergence (Actually, problems may arise even with fairly simple models.). Therefore, it is often better to specify that transformation explicitly by `logit.p[i] <- log(p[i] / (1 − p[i]))`, `p[i] <- exp(logit.p[i]) / (1 + exp(logit.p[i]))` or `p[i] <- 1 / (1 + exp(-logit.p[i]))`.

11. *Stabilized logit*: To avoid numerical overflow or underflow, you may "stabilize" the logit function by excluding extreme values (B. Wintle, pers. comm.). Here's a sketch of how to do that. The Gibbs sampling will typically get slower, but at least WinBUGS will be less likely to crash:

```
logit(psi.lim[i]) <- lpsi.lim[i]
lpsi.lim[i] <- min(999, max(-999, lpsi[i]))
lpsi[i] <- alpha.occ + beta.occ * something[i]
```

12. *Truncated normals*: Similarly, in log- or logit-normal mixtures (which we see when introducing a normal random effect into the linear predictor of a Poisson or binomial generalized linear model), you may want to truncate the zero-mean normal distribution, e.g., at ±20 (Kéry and Royle, 2009).

13. *Tinn-R special*: Tinn-R is a popular R editor. Users of Tinn-R 2.0 (or newer) may have problems writing the text file containing the BUGS model description with the `sink()` function; Tinn-R adds to that file some gibberish that will cause WinBUGS to crash. You must then use an alternative way of writing the model file. As an example, here is a workaround for the BUGS description of the model of the mean in chapter 5 that should be compatible with Tinn-R (thanks to Wesley Hochachka):

```
modelFilename = 'model.txt'
cat("
model of the mean {

# Priors
 population.mean ~ dunif(0,5000)
 precision <- 1 / population.variance
 population.variance <- population.sd * population.sd
 population.sd ~ dunif(0,100)
```

```
# Likelihood
 for( i in 1:nobs){
    mass[i] ~ dnorm(population.mean, precision)
  }
}
",fill=TRUE, file=modelFilename)
```

An alternative solution is given by Jérôme Guélat: The "R send" functions available in Tinn-R allow sending commands into R. However the "(echo=TRUE)" versions of these functions should not be used when sending the `sink()` function and its content into R. For example: one should use "R send: selection" instead of "R send: selection (echo=TRUE)".

14. *Trial runs first*: Run very short chains first, e.g., of length 12 with a burnin of 2, just to confirm that there are no coding or other errors. Only when you are satisfied that your code works and your model does what it should, increase the chain length to get a production run.

15. *Use of native WinBUGS*: A key feature of this book is that we run WinBUGS exclusively from within program R. I believe that this is much more efficient than running native WinBUGS. However, with some complex models and/or large sets, WinBUGS will be extremely slow. This may be the one exception where it is perhaps more efficient to run WinBUGS natively. You may still prepare the analysis in R as shown in this book, but only request WinBUGS to run very short Markov chains. When you set the option DEBUG = TRUE in the function `bugs()`, then WinBUGS stays open after the requested number of iterations have been conducted. Then, you can request more iterations to be executed directly in WinBUGS (i.e., using the Update Tool; see Chapter 4). You can then incrementally increase the total chain length and monitor convergence as you go. Once convergence has been achieved, do the required additional number of iterations and save them into coda files. You must do this later, since when exiting WinBUGS, the `bugs()` function only imports back into R the (small) number of iterations that you originally requested. When you have your valuable samples of your complex model's posterior distribution in coda files, use facilities provided by R packages boa or coda to import them into R and process them (e.g., compute Brooks–Gelman–Rubin convergence tests or posterior summaries for inference about the parameters).

16. *Be flexible in your modeling*: Try out different priors, e.g., for parameters representing probabilities try a uniform(0,1), a flat normal for the logit transform, or a Beta(1,1). Sometimes, and for no obvious reason, one may work while another doesn't. Similarly, WinBUGS is very sensitive to changes in the parameterization of a model. Sometimes, one way of

writing the model may work and the other doesn't, for unknown reasons, or one works much faster than the other (which, usually, is mostly an issue for more complex models than most of those featured in this book).

17. *Scale continuous covariates*: Scaling continuous covariates, so that their range does not extend too far away on either side of zero, can greatly improve mixing of the chains and often only make convergence possible (see, e.g., Section 11.4. in this book).

18. *WinBUGS hangs after compiling*: Try a restart, and if that doesn't work, find better starting values (this is an important tip from the manual).

19. *Debugging a WinBUGS analysis 1*: If something went wrong, you need to attentively read through the entire WinBUGS log file from the top to identify the first thing that went wrong. Other errors may follow, but they may not be the actual cause of the failure.

20. *Debugging a WinBUGS analysis 2*: When something doesn't work, the simplest and best advice (see also Gelman and Hill, 2007) is to go back to a simpler version of the same model, or to a similar model, that did work, and then incrementally increase the complexity of that model until you arrive at the desired model. That is, from less complex models *sneak up* on the model you want. Indeed, when using WinBUGS, you learn to always start from the simplest version of a problem and gradually build in more complexity until you are at the level of complexity that you require.

21. *DIC problems*: Surprisingly, sometimes when getting a trap (including one with the very informative title "NIL dereference (read)"), setting the argument `DIC = FALSE` in the `bugs()` function has helped.

22. *R2WinBUGS chokes*: Sometimes the R object created by R2WinBUGS is too big for one's computer. Then, use BOA or CODA to read in the coda files directly, and use their facilities to produce your inference in this way (e.g., convergence diagnostics and posterior summaries).

23. *Identifiability/estimability of parameters*: To see whether two or more parameters are difficult to estimate separately, you can plot the values of their Markov chains against each other.

24. *Check of model adequacy*: Do residual analysis, posterior predictive checks, and cross-validation to see whether your model appears to be an adequate representation of the main features in the data.

25. *Predictions*: They are a very important part of inference: i.e., the estimation of unobserved or future data. One particularly useful way to examine predictions is to estimate what a response would look like for a chosen combination of values of the explanatory variables. The generation and examination of such predicted values is an important

method to understand complex models (for instance, to see what a particular interaction means) and also needed to illustrate the results of an analysis, e.g., as a figure in a paper.

26. *Sensitivity analysis for priors*: Consider assessing prior sensitivity, i.e., repeat your analyses, or those for key models, with different prior specifications and see whether your inference is robust in this respect. If it is not, then not all is lost, but you must report on that in the methods section of your paper.

27. *Have a healthy distrust in your solutions*: Always inspect your inference, e.g., plot predictions against observed values for quantities that can be observed, to make sure that the WinBUGS solution is sensible. Also watch out for unexplained differences in parameter estimates between neighboring models, e.g., those that differ by only one covariate or some other rather minor model characteristic. This can be an indication that something went wrong or that there are estimability problems with the model for your data set.

28. *NAs and NaNs*: When dealing with data in multidimensional arrays, a very useful R package is 'reshape'. The newer versions the reshape package in R 2.9 use an NaN to fill in NAs. This makes WinBUGS very unhappy—you must have NA, not NaN. In general, this is probably good to know about BUGS and newer versions of other packages may be doing the same thing. So, if you use the melt/cast functions in reshape to organize data, then you will need to update your code in the newer R versions by adding "fill=NA_real_". Example: Ymat=cast (data.melt, SppCode~JulianDate~GridCellID, fun.aggregate=mean, fill=NA_real_) (tipp from Beth Gardner).

29. *Long Windows addresses*: WinBUGS doesn't like too long Windows addresses (C:\My harddisk\Important stuff\Less important stuff\ …) for its working directory. Hence, you should not bury your WinBUGS analyses too much down in a tree hierarchy.

30. *VISTA problems*: Windows VISTA has caused all sorts of "challenges" in workshops taught using this book—be prepared! One problem was that the default BUGS directory is not the same as that stated in section 3.4 of the book.

References

Aitkin, M., Francis, B., Hinde, J., Darnell, R., 2009. Statistical Modelling in R. Oxford University Press, Oxford, UK.

Altwegg, R., Wheeler, M., Erni, B., 2008. Climate and the range dynamics of species with imperfect detection. Biol. Lett. 4, 581–584.

Andrewartha, H.G., Birch, L.C., 1954. The Distribution and Abundance of Animals. University of Chicago Press, Chicago.

Araujo, M.B., Guisan, A., 2006. Five (or so) challenges for species distribution modeling. J. Biogeogr. 33, 1677–1688.

Bailey, L.L., Hines, J.E., Nichols, J.D., MacKenzie, D.I., 2007. Sampling design trade-offs in occupancy studies with imperfect detection: examples and software. Ecol. Appl. 17, 281–290.

Bernardo, J.M., 2003. Bayesian Statistics. Encyclopedia of Life Support Systems (EOLSS). Probability and Statistics. UNESCO, Oxford, UK. <http://www.uv.es/~bernardo/BayesStat2.pdf> (accessed 4 May 2009).

Bolker, B.M., 2008. Ecological Models and Data in R. Princeton University Press, NJ.

Borchers, D.L., Buckland, S.T., Zucchini, W., 2002. Estimating Animal Abundance. Springer, London.

Borchers, D.L., Efford, M.G., 2008. Spatially explicit maximum likelihood methods for capture-recapture studies. Biometrics 64, 377–385.

Brooks, S.P., 2003. Bayesian computation: a statistical revolution. Philos. Trans. R. Soc. Lond. A. 361, 2681–2697.

Buckland, S.T., Anderson, D.R., Burnham, K.P., Laake, J.L., Borchers, D.L., Thomas, L., 2001. Introduction to Distance Sampling. Oxford University Press, Oxford.

Burnham, K.P., Anderson, D.R., 2002. Model Selection and Multimodel Inference: Practical Information-Theoretic Approach. Springer, New York.

Clark, J.S., 2007. Models for Ecological Data: An Introduction. Princeton University Press, NJ.

Clark, J.S., Ferraz, G., Oguge, N., Hays, H., DiCostanzo, J., 2005. Hierarchical Bayes for structured, variable populations: from recapture data to life-history prediction. Ecology 86, 2232–2244.

Crawley, M.J., 2005. Statistics. An Introduction Using R. Wiley, Chichester, West Sussex.

Dalgaard, P., 2001. Introductory Statistics with R. Springer, New York.

de Valpine, P., 2009. Shared challenges and common ground for Bayesian and classical analysis of hierarchical statistical models. Ecol. Appl. 19, 584–588.

de Valpine, P., Hastings, A., 2002. Fitting population models incorporating process noise and observation error. Ecol. Monogr. 72, 57–76.

Dellaportas, P., Forster, J.J., Ntzoufras, I., 2002. On Bayesian model and variable selection using MCMC. Stat. Comput. 12, 27–36.

Dennis, B., 1996. Discussion: should ecologists become Bayesians? Ecol. Appl. 6, 1095–1103.

Dennis, B., Ponciano, J.M., Lele, S.R., Taper, M.L., Staples, D.F., 2006. Estimating density dependence, process noise, and observation error. Ecol. Monogr. 76, 323–341.

Dodd Jr., C.K., Dorazio, R.M., 2004. Using counts to simultaneously estimate abundance and detection probabilities in salamander surveys. Herpetologica 60, 468–478.

Dorazio, R.M., 2007. On the choice of statistical models for estimating occurrence and extinction from animal surveys. Ecology 88, 2773–2782.

Efford, M.G., 2004. Density estimation in live-trapping studies. Oikos 106, 598–610.

Efford, M.G., Dawson, D.K., Borchers, D.L., 2009. Population density estimated from locations of individuals on a passive detector array. Ecology 90, 2676–2682.

Elith, J., et al. 2006. Novel methods improve prediction of species' distributions from occurrence data. Ecography 29, 129–151.

Ellison, A.M., 1996. An introduction to Bayesian inference for ecological research and environmental decision-making. Ecol. Appl. 6, 1036–1046.

Fiske, I., Chandler, R., 2010. unmarked: models for data from unmarked animals. R package version 0.9.0. http://cran.r-project.org/web/packages/unmarked/index.html (accessed 1 March 2010).

Gelfand, A.E., Schmidt, A.E., Wu, S., Silander Jr., J.A., Latimer, A., Rebelo, A.G., 2005. Modelling species diversity through species level hierarchical modelling. Appl. Stat. 54, 1–20.

Gelman, A., 2005. Analysis of variance: why it is more important than ever (with discussion). Ann. Stat. 33, 1–53.

Gelman, A., 2006. Prior distributions for variance parameters in hierarchical models. Bayesian Anal. 1, 514–534.

Gelman, A., Carlin, J.B., Stern, H.S., Rubin, D.B., 2004. Bayesian Data Analysis. Second ed. CRC/Chapman & Hall, Boca Raton, FL.

Gelman, A., Hill, J., 2007. Data Analysis Using Regression and Multilevel/Hierarchical Models. Cambridge University Press, Cambridge.

Gelman, A., Meng, X.-L., Stern, H.S., 1996. Posterior predictive assessment of model fitness via realized discrepancies (with discussion). Stat. Sin. 6, 733–807.

Geman, S., Geman, S., 1984. Stochastic relaxation, Gibbs distributions, and the Bayesian restoration of images. IEEE Trans. Pattern. Anal. Mach. Intell. 6, 721–741.

Gilks, W.R., Thomas, A., Spiegelhalter, D.J., 1994. A language and program for complex Bayesian modeling. Statistician 43, 169–178.

Gu, W., Swihart, R.K., 2004. Absent or undetected? Effects of non-detection of species occurrence on wildlife-habitat models. Biol. Conserv. 116, 195–203.

Hanski, I., 1998. Metapopulation dynamics. Nature 396, 41–49.

Hastings, W.K., 1970. Monte Carlo sampling methods using Markov chains and their applications. Biometrika 57, 97–109.

Joseph, L.N., Elkin, C., Martin, T.G., Possingham, H.P., 2009. Modeling abundance using N-mixture models: the importance of considering ecological mechanisms. Ecol. Appl. 19, 631–642.

Kadane, J.B., Lazar, N.A., 2004. Methods and criteria for model selection. J. Am. Stat. Assoc. 99, 279–290.

Kéry, M., 2002. Inferring the absence of a species – A case study of snakes. J. Wildl. Manage 66, 330–338.

Kéry, M., 2004. Extinction rate estimates for plant populations in revisitation studies: the importance of detectability. Conserv. Biol. 18, 570–574.

Kéry, M., 2008. Estimating abundance from bird counts: binomial mixture models uncover complex covariate relationships. Auk 125, 336–345.

Kéry, M., Dorazio, R.M., Soldaat, L., van Strien, A., Zuiderwijk, A., Royle, J.A., 2009. Trend estimation in populations with imperfect detection. J. Appl. Ecol. 46, 1163–1172.

Kéry, M., Gardner, B., Monnerat, C., 2010b. Predicting species distributions from checklist data using site-occupancy models. J. Biogeogr. (in press)

Kéry, M., Juillerat, L., 2004. Sex ratio estimation and survival analysis for Orthetrum coerulescens (Odonata, Libellulidae). Can. J. Zool. 82, 399–406.

Kéry, M., and Royle, J.A., 2009. Inference about species richness and community structure using species-specific occupancy models in the national Swiss breeding bird survey MHB. pp. 639–656 in Thomson, D.L., Cooch, E.G., and Conroy, M.J. (eds.) Modeling

Demographic Processes in Marked Populations. Series: Environmental and Ecological Statistics, Vol. 3, Springer.

Kéry, M., Royle, J.A., 2010. Hierarchical modeling and estimation of abundance in metapopulation designs. J. Anim. Ecol. 79, 453–461.

Kéry, M., Royle, J.A., Schmid, H., 2005. Modeling avian abundance from replicated counts using binomial mixture models. Ecol. Appl. 15, 1450–1461.

Kéry, M., Royle, J.A., Schmid, H., Schaub, M., Volet, B., Häfliger, G., Zbinden, N., 2010a. Site-occupancy distribution modeling to correct population-trend estimates derived from opportunistic observations. Conserv. Biol. (in press).

Kéry, M., Schmidt, B.R., 2008. Imperfect detection and its consequences for monitoring for conservation. Community Ecol. 9, 207–216.

Kéry, M., Spillmann, J.H., Truong, C., Holderegger, R., 2006. How biased are estimates of extinction probability in revisitation studies? J. Ecol. 94, 980–986.

King, R., Morgan, B.J.T., Gimenez, O., Brooks, S.P., 2009. Bayesian Analysis for Population Ecology. Chapman and Hall/CRC, Boca Raton, FL.

Krebs, C.J., 2001. Ecology: The Experimental Analysis of Distribution and Abundance. Fifth ed. Benjamin Cummings, Menlo Park, CA.

Kuo, L., Mallick, B., 1998. Variable selection for regression models. Sankhya 60B, 65–81.

Lambert, P.C., Sutton, A.J., Burton, P.R., Abrams, K.R., Jones, D.R., 2005. How vague is vague? A simulation study of the impact of the use of vague prior distributions in MCMC using WinBUGS. Stat. Med. 24, 2401–2428.

Latimer, A.M., Wu, S., Gelfand, A.E., Silander Jr., J.A., 2006. Building statistical models to analyse species distributions. Ecol. Appl. 16, 33–50.

Le Cam, L., 1990. Maximum likelihood: an introduction. Int. Stat. Rev. 58, 153–171.

Lee, Y., Nelder, J.A., 2006. Double hierarchical generalised linear models. Appl. Stat. 55, 139–185.

Lee, Y., Nelder, J.A., Pawitan, Y., 2006. Generalized Linear Models with Random Effects. Unified Analysis via H-likelihood. Chapman and Hall/CRC, Boca Raton, FL.

Lele, S.R., Dennis, B., 2009. Bayesian methods for hierarchical models: are ecologists making a Faustian bargain? Ecol. Appl. 19, 581–584.

Lindley, D.V., 1983. Theory and practice of Bayesian statistics. Statistician 32, 1–11.

Lindley, D.V., 2006. Understanding Uncertainty. Wiley, Hoboken, NJ.

Link, W.A., Barker, R.J., 2006. Model weights and the foundations of multimodel inference. Ecology 87, 2626–2635.

Link, W.A., Barker, R.J., 2010. Bayesian Inference with Ecological Examples. Academic Press, San Diego, CA.

Link, W.A., Cam, E., Nichols, J.D., Cooch, E.G., 2002. Of BUGS and birds: Markov chain Monte Carlo for hierarchical modeling in wildlife research. J. Wildl. Manage. 66, 277–291.

Link, W.A., Sauer, J.R., 2002. A hierarchical analysis of population change with application to Cerulean warblers. Ecology 83, 2832–2840.

Littell, R.C., Milliken, G.A., Stroup, W.W., Wolfinger, R.D., Schabenberger, O., 2008. SAS for Mixed Models. Second ed. SAS Institute, Cary, NC.

Little, R.J.A., 2006. Calibrated Bayes: a Bayes/Frequentist roadmap. Am. Stat. 60, 213–223.

Lunn, D., Spiegelhalter, D., Thomas, A., Best, N., 2009. The BUGS project: evolution, critique and future directions. Stat. Med. 28, 3049–3067.

MacKenzie, D.I., 2006. Modeling the probability of resource use: the effect of, and dealing with, detecting a species imperfectly. J. Wildl. Manage. 70, 367–374.

MacKenzie, D.I., Kendall, W.L., 2002. How should detection probability be incorporated into estimates of relative abundance? Ecology 83, 2387–2393

MacKenzie, D.I., Nichols, J.D., Hines, J.E., Knutson, M.G., Franklin, A.B., 2003. Estimating site occupancy, colonization and local extinction when a species is detected imperfectly. Ecology 84, 2200–2207.

MacKenzie, D.I., Nichols, J.D., Lachman, G.B., Droege, S., Royle, J.A., Langtimm, C.A., 2002. Estimating site occupancy rates when detection probability rates are less than one. Ecology 83, 2248–2255.

MacKenzie, D.I., Nichols, J.D., Royle, J.A., Pollock, K.H., Hines, J.E., Bailey, L.L., 2006. Occupancy Estimation and Modeling: Inferring Patterns and Dynamics of Species Occurrence. Elsevier, San Diego, CA.

MacKenzie, D.I., Royle, J.A., 2005. Designing occupancy studies: general advice and allocating survey effort. J. Appl. Ecol. 42, 1105–1114.

Martin, T.G., Kuhnert, P.M., Mergersen, K., Possingham, H.P., 2005. The power of expert opinion in ecological models using Bayesian methods: impact of grazing on birds. Ecol. Appl. 15, 266–280.

Mazzetta, C., Brooks, S., Freeman, S.N., 2007. On smoothing trends in population index modeling. Biometrics 63, 1007–1014.

McCarthy, M.A., 2007. Bayesian Methods for Ecology. Cambridge University Press, Cambridge.

McCarthy, M.A., Masters, P., 2005. Profiting from prior information in Bayesian analyses of ecological data. J. Appl. Ecol. 42, 1012–1019.

McCullagh, P., Nelder, J.A., 1989. Generalized Linear Models. Second ed. Chapman and Hall, London.

McCulloch, C.E., Searle, S.R., 2001. Generalized, Linear, and Mixed Models. Wiley, New York.

Metropolis, N., Rosenbluth, A.W., Rosenbluth, M.N., Teller, A.H., Teller, E., 1953. Equation of state calculations by fast computing machines. J. Chem. Phys. 21, 1087–1092.

Millar, R.B., 2009. Comparison of hierarchical Bayesian models for over-dispersed count data using DIC and Bayes factors. Biometrics 65, 962–969.

Monneret, R.-J., 2006. Le faucon pèlerin. Delachaux and Niestlé, Paris, France.

Nelder, J.A., Wedderburn, R.W.M., 1972. Generalized linear models. J. R. Stat. Soc. Ser. A. 135, 370–384.

Ntzoufras, I., 2009. Bayesian Modeling Using WinBUGS. Wiley, Hoboken, NJ.

Pellet, J., Schmidt, B.R., 2005. Monitoring distributions using call surveys: estimating site occupancy, detection probabilities and inferring absence. Biol. Conserv. 123, 27–35.

Pinheiro, J.C., Bates, D.M., 2000. Mixed-Effects Models in S and S-Plus. Springer, New York.

Ponciano, J.M., Taper, M.L., Dennis, B., Lele, S.R., 2009. Hierarchical models in ecology: confidence intervals, hypothesis testing, and model selection using data cloning. Ecology 90, 356–362.

Powell, L.A., 2007. Approximating variance of demographic parameters using the delta method: a reference for avian biologists. Condor 109, 949–954.

R Development Core Team 2007. R: A language and Environment for Statistical Computing. R Foundation for Statistical Computing, Vienna, Austria. <http://www.R-project.org> (accessed 4 May 2009).

Robinson, G.K., 1991. That BLUP is a good thing: the estimation of random effects (with discussion). Stat. Sci. 6, 15–51.

Royle, J.A., 2004a. N-mixture models for estimating population size from spatially replicated counts. Biometrics 60, 108–115.

Royle, J.A., 2004b. Generalized estimators of avian abundance from count survey data. Anim. Biodivers. Conserv. 27, 375–386.

Royle, J.A., 2008. Modeling individual effects in the Cormack-Jolly-Seber model: a state-space formulation. Biometrics 64, 364–370.

Royle, J.A., Dorazio, R.M., 2006. Hierarchical models of animal abundance and occurrence. J. Agric. Biol. Environ. Stat. 11, 249–263.

Royle, J.A., Dorazio, R.M., 2008. Hierarchical Modeling and Inference in Ecology. Academic Press, Amsterdam.

Royle, J.A., Karanth, K.U., Gopalaswamy, A.M., Kumar, N.S., 2009. Bayesian inference in camera trapping studies for a class of spatial capture-recapture models. Ecology 90, 3233–3244.

Royle, J.A., Kéry, M., 2007. A Bayesian state-space formulation of dynamic occupancy models. Ecology 88, 1813–1823.

Royle, J.A., Kéry, M., Gautier, R., Schmid, H., 2007. Hierarchical spatial models of abundance and occurrence from imperfect survey data. Ecol. Monogr. 77, 465–481.

Royle, J.A., Nichols, J.D., 2003. Estimating abundance from repeated presence-absence data or point counts. Ecology 84, 777–790.

Royle, J.A., Nichols, J.D., Kéry, M., 2005. Modeling occurrence and abundance of species with imperfect detection. Oikos 110, 353–359.

Royle, J.A., Young, K.G., 2008. A hierarchical model for spatial capture-recapture data. Ecology 89, 2281–2289.

Sauer, J.R., Link, W.A., 2002. Hierarchical modeling of population stability and species group attributes from survey data. Ecology 86, 1743–1751.

Schmidt, B.R., 2005. Monitoring the distribution of pond-breeding amphibians when species are detected imperfectly. Aquat. Conserv. Mar. Freshwater Ecosyst. 15, 681–692.

Schmidt, B.R., 2008. Neue statistische Verfahren zur Analyse von Monitoring- und Verbreitungsdaten von Amphibien und Reptilien. Zeitschrift für Feldherpetologie 15, 1–14.

Smith, A.F.M., Gelfand, A.E., 1992. Bayesian statistics without tears: A sampling-resampling perspective. Am. Stat. 46, 84–88.

Spiegelhalter, D., Best, N.G., Carlin, B.P., van der Linde, A., 2002. Bayesian measures of complexity and fit (with discussion). J. R. Stat. Soc. Series B. 64, 583–639.

Spiegelhalter, D., Thomas, A., Best, N.G., 2003. WinBUGS User Manual, Version 1.4. MCR Biostatistics Unit, Cambridge.

Sturtz, S., Ligges, U., Gelman, A., 2005. R2WinBUGS: a package for running WinBUGS from R. J. Stat. Softw. 12, 1–16.

Swain, D.P., Jonsen, I.D., Simon, J.E., Myers, R.A., 2009. Assessing threats to species at risk using stage-structured state-space models: mortality trends in skate populations. Ecol. Appl. 19, 1347–1364.

Tyre, A.J., Tenhumberg, B., Field, S.A., Niejalke, D., Parris, K., Possingham, H.P., 2003. Improving precision and reducing bias in biological surveys: Estimating false-negative error rates. Ecological Applications 13, 1790–1801.

Venables, W.N., Ripley, B.D., 2002. Modern Applied Statistics with S. Springer, New York.

Ver Hoef, J.M., Jansen, J.K., 2007. Space-time zero-inflated count models of Harbour seals. Environmetrics 18, 697–712.

Wade, P.R., 2000. Bayesian methods in conservation biology. Conserv. Biol. 14, 1308–1316.

Webster, R.A., Pollock, K.H., Simons, T.R., 2008. Bayesian spatial modeling of data from avian point count surveys. J. Agric. Biol. Environ. Stat. 13, 121–139.

Welham, S., Cullis, B., Gogel, B., Gilmour, A., Thompson, R., 2004. Prediction in linear mixed models. Aust. N. Z. J. Stat. 46, 325–347.

Wenger, S.J., Freeman, M.C., 2008. Estimating species occurrence, abundance, and detection probability using zero-inflated distributions. Ecology 89, 2953–2959.

White, G.C., Burnham, K.P., 1999. Program MARK: survival estimation from populations of marked animals. Bird Study 46 (Suppl.), 120–138.

Wikle, C.K., 2003. Hierarchical Bayesian models for predicting the spread of ecological processes. Ecology 84, 1382–1394.

Williams, B.K., Nichols, J.D., Conroy, M.J., 2002. Analysis and Management of Animal Populations. Academic Press, San Diego, CA.

Woodworth, G.G., 2004. Biostatistics. A Bayesian Introduction. Wiley, Hoboken, NJ.

Yoccoz, N.G., Nichols, J.D., Boulinier, T., 2001. Monitoring biological diversity in space and time. Trends Ecol. Evol. 16, 446–453.

Zeileis, A., Kleiber, C., Jackman, S., 2008. Regression models for count data in R. J. Stat. Softw. 27, <http://www.jstatsoft.org/v27/i08/> (accessed 4 May 2009).

Index